Metal Organic Frameworks

Related Titles

Biopolymers Based Advanced Materials
ISBN: 978-0-6482205-4-1 (e-book)
ISBN: 978-0-6482205-5-8 (hardcover)

Functional Polymer Blends and Nanocomposites
ISBN: 978-0-6482205-6-5 (e-book)
ISBN: 978-0-6482205-7-2 (hardcover)

Functional Nanomaterials and Nanotechnologies: Applications for Energy & Environment
ISBN: 978-0-6482205-2-7 (e-book)
ISBN: 978-0-6482205-3-4 (softcover)

Advances in Polymer Technology: Material Development, Properties and Performance Evaluation
ISBN: 978-1-925823-00-4 (e-book)
ISBN: 978-1-925823-01-1 (hardcover)

Polymer Nanomaterials for Specialty Applications
ISBN: 978-1-925823-03-5 (e-book)
ISBN: 978-1-925823-04-2 (hardcover)

Advanced Materials
ISBN: 978-1-925823-05-9 (e-book)
ISBN: 978-1-925823-06-6 (hardcover)

Biofuels
ISBN: 978-1-925823-12-7 (e-book)
ISBN: 978-1-925823-13-4 (hardcover)

Liquid Crystalline Polymers
ISBN: 978-1-925823-16-5 (e-book)
ISBN: 978-1-925823-17-2 (hardcover)

Polymer Nanocomposites: Emerging Applications
ISBN: 978-1-925823-14-1 (e-book)
ISBN: 978-1-925823-15-8 (hardcover)

Metal Organic Frameworks

Dr. Vikas Mittal

Editor

CWP

Central West Publishing

Disclaimer
Every effort has been made by the publisher, editor and authors while preparing this book, however, no warranties are made regarding the accuracy and completeness of the content. The publisher, editor and authors disclaim without any limitation all warranties as well as any implied warranties about sales, along with fitness of the content for a particular purpose. Citation of any website and other information sources does not mean any endorsement from the publisher and authors. For ascertaining the suitability of the contents contained herein for a particular lab or commercial use, consultation with the subject expert is needed. In addition, while using the information and methods contained herein, the practitioners and researchers need to be mindful for their own safety, along with the safety of others, including the professional parties and premises for whom they have professional responsibility. To the fullest extent of law, the publisher, editor and authors are not liable in all circumstances (special, incidental, and consequential) for any injury and/or damage to persons and property, along with any potential loss of profit and other commercial damages due to the use of any methods, products, guidelines, procedures contained in the material herein.

NATIONAL LIBRARY OF AUSTRALIA

A catalogue record for this book is available from the National Library of Australia

ISBN (print): 978-1-925823-57-8
ISBN (e-book): 978-1-925823-56-1

Contents

8.   Metal Organic Frameworks for H₂S Separation from Natural Gas   **173**

Gigi George

9.   Metal Organic Frameworks for Detection of Organic Nitro Compounds   **187**

Yadagiri Rachuri, Bhavesh Parmar, Jintu Francis Kurisingal, Kamal Kumar Bisht, Dae-Won Park and Eringathodi Suresh

sensor material for the detection of organic nitro aliphatic and aromatic compounds are the highlighted in Chapter 9.

The book would not have been successfully accomplished without the support of chapter contributors. The book is dedicated to my family for unswerving support, constant motivation as well as constructive suggestions for improvement.

Vikas MITTAL

Antimicrobial Metal Organic Frameworks

Ana Arenas Vivo and Patricia Horcajada*

Advanced Porous Materials Unit, IMDEA Energy Institute, Av. Ramón de la Sagra 3, 28935 Móstoles-Madrid, Spain

Corresponding author: patricia.horcajada@imdea.org

1.1 Introduction

Bacteria are the most common microorganisms present on Earth and can be found in almost every habitat. While some of these bacteria are beneficial, a great proportion is the cause of the propagation of several diseases and infections [1]. Apart from the direct health care implications, unwanted bacteria colonization is of high relevance in other less obvious industries and applications (food packaging, marine shipping, water treatment, heat exchanging systems, etc.), thus, having a significant environmental and socioeconomic impact [2].

Over the last decades, there has been a radical change in the approach to treat infections. Although infectious diseases have been traditionally treated with antibiotics, their frequent use has led to the prevalence of resistant bacterial strains [3]. While classical evolving infections are believed to entail the free-floating (planktonic) bacteria, the appearance of chronic infections is related with sessile aggregation of microorganisms, known as biofilms [4]. These bacterial micro-environments are formed by irreversibly attached microorganisms and extracellular polymeric substances (EPS) in interaction with a substrate. The biofilm formation process involves the development of a protective environment to the growing colony ideal for bacterial proliferation. It is well known that cells within biofilms experience: i) differentiation in response to the local conditions they are subjected to, ii) cell-to-cell communication (quorum-sensing) and iii) demonstrated resistance to environmental stresses [5]. Therefore, the need for new bactericides is a pressing real chal-

Metal Organic Frameworks, edited by Vikas Mittal
© 2019 Central West Publishing, Australia

lenge and encourages biologists, chemists and materials scientists to develop efficient antibacterial agents. Different strategies have been developed in order to resolve the issue during several stages of bio-film formation (*i.e.* contact, attachment, proliferation, maturation and dispersion), such as 1) limiting switch from planktonic to bio-film lifestyle, 2) limiting initial adhesion and interaction, 3) interfering in bacterial communication, with quorum-sensing autoinducers, 4) developing anti-adhesive surfaces and 5) promoting dispersion [6]. Several chemical disinfectants, both inorganic (TiO_2, ZnO, Ag, NO, I, etc.) and inorganic-organic hybrid materials (montmorillonite, zeolites and silica based sol-gels) [7] have been increasingly used against Gram (-) and Gram (+) bacteria in several consumer products (e.g. catheters, prosthesis, food wrap films, etc.) [8,9].

In recent years, a new type of hybrid materials, called metal organic frameworks (MOFs) (also known as porous coordination polymers (PCPs)), has been developed as promising biocide materials [10]. These crystalline solids consist of inorganic units (e.g. atoms, clusters and chains) connected *via* ionocovalent bonding to polydentate organic ligands (e.g. carboxylates, phosphonates and azolates), leading to porous 3D frameworks (Figure 1.1). Due to their exceptional porosity (up to Brunauer-Emmett-Teller surfaces-S_{BET} = 7000 m^2 g^{-1}; pore diameter = 3-90 Å) [11] and compositional and structural versatility, MOFs stand out as ideal candidates for several strategic applications, both at industrial and social level (e.g. separation, sensing, catalysis, energy, etc.) [12]. More recently, their use has

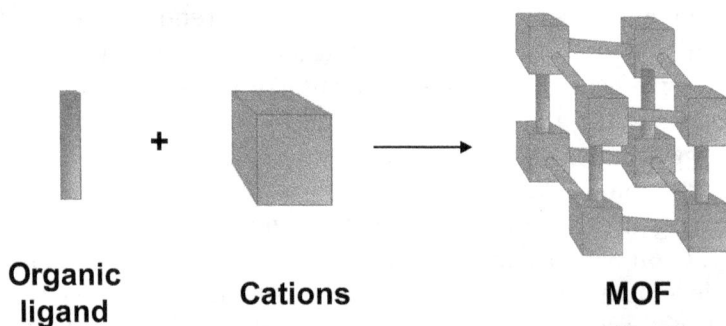

Organic ligand   **Cations**   **MOF**

Figure 1.1 Schematic representation of a MOF.

been explored in biomedicine (e.g. diagnosis, drug delivery and cosmetics) [13]. Some of the advantages of MOFs compared to other antibacterial materials are as follows: i) both organic and inorganic components of MOFs can provide bactericidal activity by the generation of reactive oxygen species (ROS), ii) they have a uniform and ordered distribution of active sites and iii) metal release to the media during degradation tends to be more homogeneous. In addition, not only their framework can be bioactive, but they can also entrap diverse bactericidal agents and increase their disinfection power [14].

Taking into consideration these properties, MOFs as biocide agents can be classified according to the origin of their bioactivity: i) MOFs with intrinsic antimicrobial activity, which comes from the network, or ii) MOFs with extrinsic antimicrobial activity, which arises from a hosting therapeutic agent that is associated to their structure [15]. The first one entails the controlled degradation of the MOF for the liberation of the active components (metal, ligand or both) and these structures are normally referred to as 'ion reservoirs' [16,17]. The second classification requires the release of the active ingredients (AIs, e.g. metallic nanoparticles, NPs, drugs, etc.) *via* diffusion through the MOF porosity or leakage upon MOF dissolution. Since the first reported one dimensional (1D) PCP with biocide properties in 2010-2012 [18,21], diverse studies have explored the use of MOFs as antibacterial agents. Recently, Wyszogrodzka *et al.* [14] also reviewed the topic in greater details.

The aim of this book chapter is to review the development of antimicrobial MOFs (limiting them to porous 3D structures), with either intrinsic or extrinsic activity, including not only some AI immobilization strategies and examples of their integration in several devices (membranes, thin films, etc.), but also their applications.

1.2 Intrinsic Antimicrobial Activity

As mentioned above, the origin of the bactericide effect in MOFs with intrinsic antimicrobial activity relies on the activity of its individual (organic and/or inorganic) components. It is well-known that certain metals (e.g. Cu, Ti, Ag, Zn and Fe) [22-24] possess a bioactive character. As a consequence, these have been selected in several studies as inorganic building units for the formulation of MOFs to obtain antibacterial frameworks. Other strategy is the use of drugs and other therapeutic biomolecules as organic ligands for building metal biomolecules frameworks (or bioMOFs). Some of these 'bioactive ingredients' have been further discussed by Rojas *et al.* [17]. Even more, both the cation and ligand can have disinfectant properties, and, therefore, the final MOF exhibits an additive or synergistic bactericide effect.

Some of the advantages of intrinsic antimicrobial activity of MOFs over other solids are: i) active antibacterial agents are obtained relatively easily, by just considering the MOF synthesis protocol (e.g. without the need of other impregnation steps), ii) there is no need to achieve high porosity in these frameworks as the bactericidal activity comes from the constituents, iii) they can be synthesized with different particle sizes, which strongly determine their final biodistribution, toxicity and activity [25], iv) they can remain stable in bio-relevant conditions with time to achieve their purpose and v) they have an uniform and ordered distribution of active sites, along with more homogeneous release to the media during degradation.

In this section, 3D MOFs with intrinsic biocidal activity are compiled, organized according to the source of their effect: derived from the metal, ligand or both.

1.2.1 Antimicrobial Activity Derived from Metal

The MOFs studied as 'metal reservoirs' were firstly compiled by Wyszogrodzka *et al.* [14] in 2016 (Table 1.1). In combination with these, an up-to-date compilation is presented here, organized by the nature of the selected metallic cation.

The antibacterial action of silver has been extensively utilized, from consumer products to medical devices. Although its mechanism of action is not completely understood, it is believed that it results from the Ag^+ delivery, which interacts with the bacterial cell membrane and plasma components, thus, affecting their function [22]. Interestingly, Liu *et al.* [26] were the first to report the synthesis of three Ag-based metal organoboron frameworks for use as Ag^+ reservoir against planktonic *Escherichia coli* (facultative aerobic Gram (-)) and *Staphylococcus aureus* (aerobic Gram (+)). The as-prepared micrometric crystals were subjected to the standard disk diffusion method, thus, measuring the bacterial growth inhibition zone to evaluate the biocidal activity. The minimum inhibitory concentration (MIC) for both bacteria was 300 ppm of MOF, Ag content being 12 wt% (36 ppm). The Ag^+ release from the three MOF structures, determined in distilled water by inductively coupled plasma optical emission spectroscopy (ICP-OES), was 3.7, 4.4 and 5.2 10^{-6} g per day. Berchel *et al.* [16] also evaluated the antibacterial activity of the Ag$_3$[C$_7$H$_4$O$_5$P] MOF, previously reported by the same group [27], against *S. aureus, E. coli* and *Pseudomonas aureginosa* (aerobic Gram (-)). The minimal active bactericidal concentration (MBC) in the culture medium was determined by means of optical density and plate count, with Ag$_3$[C$_7$H$_4$O$_5$P] displaying interesting bactericidal properties with a broad antibacterial spectrum. Given the proven negligible bactericidal effect of the ligand, 3-phosphonobenzonic acid, the authors concluded that the MOF activity might exclusively come from the Ag$^+$ released from the MOF.

To further develop new MOFs based on Ag and aromatic–carboxylic acids ligands containing hydroxyl and pyridyl groups, Lu *et al.* [28] generated three new structures, Ag$_2$(3-NPTA)(bipy)$_{0.5}$(H$_2$O) [28] (Figure 1.2), Ag$_2$[(O-IPA)(H$_2$O)·(H$_3$O)] and Ag$_5$[(PYDC)$_2$(OH)] [29], with the ligands 3-/4-nitrophthalic acid and 4,4'-bipyridyl in the first, whereas 5-hydroxyisophthalic acid and pyridine-3,5-dicarboxylic acid in the other two. The developed MOFs exhibited a significant antimicrobial activity against *S. aureus*

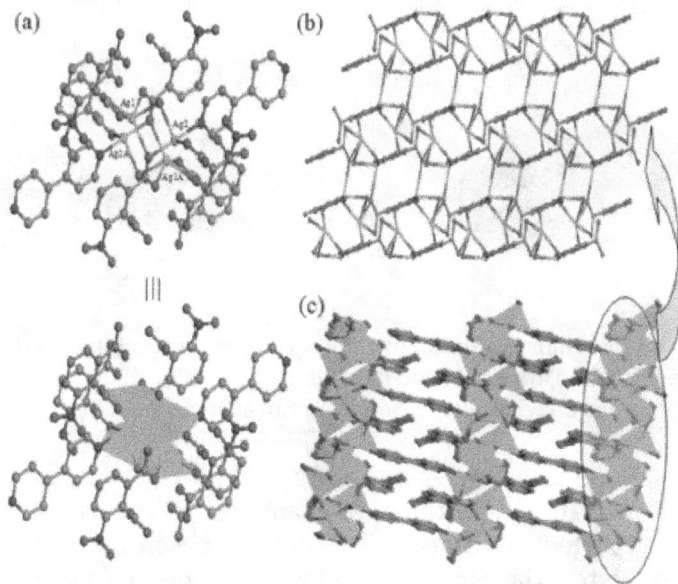

Figure 1.2 Ball and stick representation as well as polyhedral representation of the tetranuclear unit (a), a 2-D layer magnification part of the structure (b) and 3D framework (c) in Ag$_2$(3-NPTA)(bipy)$_{0.5}$(H$_2$O). Reproduced from Reference 28 with permission from Royal Society of Chemistry.

and *E. coli*. Their MICs, estimated by optical density, were in the range of 10-20 ppm of MOF, which indicated relatively higher antibacterial activity in suspension as compared to other MOFs (20 *vs.* 300 ppm [26]). These Ag-based MOFs also showed higher activity in the inhibition zone test as compared to Ag nanoparticles (AgNPs). Based on the findings from ICP-OES and transmission electron microscopy (TEM) images, the authors proposed that the bactericidal mechanism was due to the delivery of Ag^+ and their interference with the cell membrane, resulting in the loss of cellular cohesion (Figure 1.3).

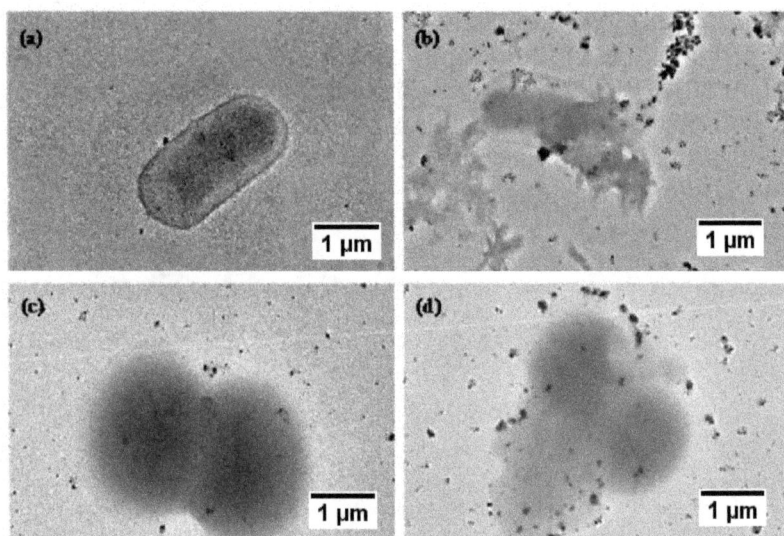

Figure 1.3 TEM morphological images of cell structures: intact and damaged E. coli (a, b); intact and damaged S. aureus (c, d). Reproduced from Reference 28 with permission from Royal Society of Chemistry.

For the development of novel Ag-MOFs based on polyoxometalates (P_2W_{18}), useful for their abundant coordination sites, Wang *et al.* [30] reported three original structures based on Ag as cation. They evaluated their potential biocide properties *via* inhibition zone, observing that the ligands (1,3-bis(1,2,4-triazol-1-y1)propane and 1,4-bis(1,2,4-triazol-1-y1)butane), dimethyl sulfoxide (solvent used for the synthesis) and P_2W_{18} displayed almost no inhibition against *S. aureus* and *E. coli*, while the Ag-MOFs exhibited a similar inhibition as $AgNO_3$ (positive blank used as reference of Ag^+ release). Zhang *et al.* [31] recently essayed the *S. aureus* viability against the SD/Ag14 open MOF, with cyclopropylacetylene and *p*-toluenesulfonate as building blocks shaping the Ag_{14} cluster. The authors determined a MIC of 5 ppm *via* plate count, which was smaller than the 20 ppm reported for AgMOFs by Lu *et al.* [28,29].

Another group of MOFs with metal-induced intrinsic antimicrobial activity are Cu-based MOFs. Again, the release of Cu^{2+} is the key to promote disinfection. These cations can adhere to bacterial cell membranes and cause their cytolysis and release of intracellular content, resulting in the loss of bacteria viability [32]. With the aim of investigating the effect of MOF morphology on bacterial viability, Wang *et al.* [33] synthesized the Cu-pyridine MOF with different shapes (rhombus layers, rhombus discs, rhombus lumps and bread-like) and determined the MIC against Gram (+) (*Bacillus subtilis* and *S. aureus*) and Gram (-) bacteria (*Salmonella enteriditis, E. coli, Proteus vulgaris* and *P. aeruginosa*). Overall, the rhombus lumps exhibited the best antibacterial performance (e.g. *E.coli* MIC = 6.25 ppm), although the mechanism of action was not analyzed. The antibacterial activity of the well-known microporous Cu trimesate (CuBTC, also called HKUST-1) [34] was first proved by Abbasi *et al.* [35] on dense coating on silk fibers (see Section: Devices and Applications). Soon afterwards, Chiericatti *et al.* [36] reported the first use of HKUST-1 as an antifungal agent. The material showed higher antifungal activity against *Saccharomyces cerevisae* than *Geotrichum candium* in plate count. The physicochemical characterizations performed after assays determined that HKUST-1 activity resulted from the release of Cu ions upon framework degradation. HKUST-1 fungicidal effect has been further studied recently by Celis-Arias *et al.* [37] against *Aspergillus niger, Fusarium solani* and *Penicillium chrysogenum*, taking into consideration the effect of the MOF particle size. HKUST-1 doped with CuO NPs (2 wt%, HKUST-1-NP) exhibited higher disinfection at lower concentration, particularly against *P. Chrysogenum*, when compared with pristine HKUST-1 and a proportional amount of CuO-NPs. HKUST-1 coated by activated carbon (AC-HKUST-1), generated by an *in-situ* ultrasound assisted synthesis, was also explored as a dye absorbent for water treatment as well as an antibacterial agent against challenging multidrug-resistant *S. aureus* and *P. aeruginosa* by Azad *et al.* [38]. The obtained MIC values were 100 and 50 ppm, respectively, without significant differences in the biocidal activity of the AC-HKUST-1 and blank HKUST-1, thus, ruling out the effect of the AC on the bactericidal effect. Nevertheless, the coated material exhibited higher dye adsorption capacity with a 97% removal as compared to 63% for HKUST-1. Additionally, the copper terephthalate surface-anchored MOF Cu-SURMOF-2 has also been proposed by Sancet *et al.* [39] to treat biofilm formation in the maritime environments against the Gram (-) halophilic *Cobetia marina*. Preliminary stability studies of the layer-by-layer synthesized MOF, carried out with X-ray diffraction (XRD) and ICP-OES, revealed that Cu–SURMOF-2 was stable in water and artificial sea water during the 2 h contact time of study. Nevertheless, the MOF lost its cohesion with *C. marina* present in the medium and had a fast release of 2 ppm of Cu in 2 h. The authors suggested that the EPS delivery affected the MOF integrity after bacterial attachment to the surface, thus, leaching Cu^+ ions that interfere with the viability of bacteria. Indeed, LIVE/DEAD staining showed 80% of *C. marina* yellow, which indicated initial viability affection. The atomic force microscopy (AFM) and scanning electron microscopy (SEM) micrographs revealed a wrinkled bacterial surface, when in contact with the Cu-SURMOF-2 for 2 h. Figure 1.4 also shows the inhibition zone test analysis.

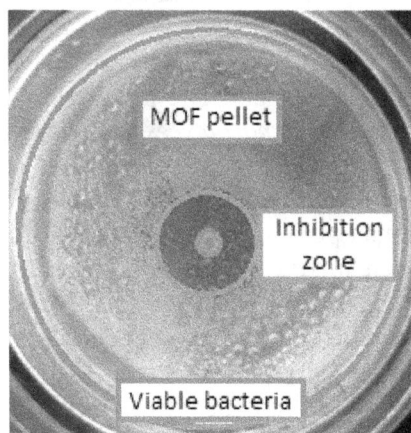

Figure 1.4 Inhibition zone test carried out with a MOF pellet (green circle) on top of a petri dish (inhibition zone (orange circle) and viable bacteria (blue circle)).

In addition, several cobalt complexes have been previously reported with proven antiviral and antibacterial efficacy [40]. In this case, it is believed that these complexes promote the ROS generation when cations are in the culture broth, thus, affecting the function of the plasma membrane proteins. Hence, it might be reasonable to use Co as a building unit of biocide MOFs. However, one might also consider the potential toxicity of cobalt when compared with, for instance, an endogenous metal such as zinc (oral LD_{50} (rats; considering sulfate salts) = 1710 and 424 mg/kg [41,42] and daily doses = 15 and 0.03 mg, for Zn and Co, respectively) [43]. In this respect, Zhuang *et al.* [44] synthesized the porous Co-TDM MOF with Co and an octa-topic carboxylate ligand, tetrakis[(3,5-dicarboxyphenyl)-oxamethyl] methane, to probe its bactericidal effect against two strains of *E. coli.* Bacterial growth was evidenced *via* optical density at 600 nm (OD_{600}) in contact with 10-15 ppm of Co-TDM over an 8 h period. This bactericidal effect was observed even after 20 minutes in contact with the *E. coli* strains. TEM observations of the damaged bacteria led to the conclusion that the cell death resulted due to the disruption of the cell membrane when in contact with the MOF. Aguado *et al.* [45] selected two cobalt imidazolates, ZIF-67 (ZIF: zeolitic imidazolate frameworks) and Co-SIM-1 (SIM: substituted imidazolate material) as bactericidal materials. The authors conducted agar plate diffusion tests and determined MIC *via* OD measurements against *S. cerevisiae, Pseudomonas putida* and *E. coli,* using AgTAZ PCP for comparison purposes, selected for its commercially available ligand (1,2,4-triazole) and the well-known activity of Ag. With a progressive cobalt release (over 3 months) within the range of previously studied Ag-based materials [16,26], the selected Co-based MOFs showed higher activity than AgTAZ. In another study, the authors dispersed Co-SIM-1 in polylactic acid (PLA), a polymer derived from renewable resources with natural hydrophilicity that reduces biofuling tendency, and electrospun into fibers with antimicrobial effect against *P. putida* and *S. aureus* [46] (see Section: Devices and Applications).

Lastly, MOFs constituted by Zn building units have also been reported as biocide materials. Zn, as a biocompatible endogenous cation (daily recommended dose = 15 mg) [43], has attracted the interest of researchers due to its bactericidal and bacteriostatic properties. The mechanism of action has been thoroughly studied and attributed to ROS generation and enhanced membrane permeability, internalization of NPs due to loss of proton motive force and uptake of toxic dissolved zinc ions [47]. Although most of the reported studies are based on the activity of extrinsic AI (see Section: Extrinsic Antimicrobial Activity), a few studies deal with the intrinsic activity of zinc. For instance, Martin-Betancor *et al.* [48] evaluated the bactericidal activity of the zeolite imidazolate framework, Zn-SIM-1, against suspensions of photosynthetic cyanobacteria (*Anabaena, Synechoccocus*) in comparison with Co-SIM-1 and AgTAZ. The biocidal activity of the materials was determined by measuring chlorophyll *a* concentration and performing inhibition zone tests. Co-SIM-1 exhibited slightly superior activity as compared to its Zn MOF analogue. As it was elucidated that the biocidal activity was a result of the dissolved metals present in the culture medium, total ion

Table 1.1 Compilation of the MOFs used as ion reservoir for bactericide applications

Compound	Metal	Ligand	Ref.
[(AgL)NO$_3$]·2H$_2$O [(AgL)CF$_3$SO$_3$]·2H$_2$O [(AgL)ClO$_4$]·2H$_2$O	Ag	L=Tris-(4-pyridylduryl)borane	[26]
Ag$_3$[C$_7$H$_4$O$_5$P]	Ag	3-Phosphonobenzoate	[16]
[Ag$_2$(O-IPA)(H$_2$O)·(H$_3$O)]	Ag	5-Hydroxyisophthalic	[29]
[Ag$_5$(PYDC)$_2$(OH)]	Ag	Pyridine-3,5-dicarboxylic acid	[29]
C$_{24}$H$_{39}$Ag$_7$N$_{24}$O$_{65}$P$_2$W$_{18}$ C$_{35}$H$_{53}$Ag$_7$N$_{30}$O$_{63}$P$_2$W$_{18}$ C$_{36}$H$_{60}$Ag$_4$N$_{27}$O$_{64}$P$_2$W$_{18}$	Ag	Polyoxometalate: P$_2$W$_{18}$O$_{62}^{6-}$ Ligand: bis(triazole) with different spacer lengths	[30]
C$_{171}$H$_{173}$Ag$_{42}$Cl$_9$O$_{10}$S$_3$	Ag	p-toluenesulfonate	[31]
CuC$_{12}$H$_{14}$N$_2$O$_8$	Cu	4,4'-dicarboxy-2,2'-bipyridine	[33]
HKUST-1	Cu	Bencene 1,3,5-tricarboxylic acid	[36]
Cu-SURMOF 2	Cu	Benzene 1,4-dicarboxylic acid	[39]
Co$_4$(H2O)$_2$(TDM)(H2O)$_8$	Co	tetrakis [(3,5-dicarboxyphenyl)oxamethyl] methane acid	[44]
ZIF-67	Co	2-methylimidazolate	[45]
Co-SIM-1	Co	4-methyl-5-imidazole carboxaldehyde	[45] [48]
AgTAZ	Ag	1,2,4-triazole	[45] [48]
Zn-SIM-1	Zn	4-methyl-5-imidazole carboxaldehyde	[48]
BioMIL-5	Zn	Azelaic acid	[49]
	Ag	Bencene 1,3,5-tricarboxylic acid	[50]

concentration was determined with two experimental techniques (ICP-OES and heavy metal bioreporter) and *via* chemical computational modelling (Visual MINTEQ). The experiments revealed a higher amount of Co dissolved in the medium, occasionally double of the Zn concentration, therefore, explaining the higher activity of the Co-SIM-1 material as compared to Zn-SIM-1.

1.2.2 Antimicrobial Activity Derived from a Cation-Ligand Combined Effect

There are plenty of references in the literature related to the MOFs made of endogenous and/or therapeutically active ligands (also known as bioMOFs) and their applications in biomedicine (drug delivery, imaging and sensing [13,15,17,25,51,52]), however, only a few reports can be found concerning their antibacterial activity.

Occasionally, the combination of different species is required to provide a cooperative therapeutic effect. BioMOFs based on antibacterial cations and bioactive ligands (Figure 1.5) represent this useful combination, as these can be specifically designed for a particular application. Such is the case of the $[Ag(\mu_3\text{-PTA=S})]_n(NO_3)_n \cdot nH_2O$ bioMOF [53], which combines the antimicrobial activity of Ag and aminophosphine 1,3,5-triaza-7-phosphaadamantane derivative (PTA=S). The bioMOF was screened against *E. coli*, *P. aeruginosa* and *S. aureus*, determining MIC values of 11, 14 and 56 ppm, respectively. A control with the corresponding amount of silver exhibited MIC values of 37, 368 and 368 ppm, whereas the MIC of the PTA=S ligand alone was over 600 ppm for the three bacteria. This led to the conclusion that the anti-bactericidal effect of $[Ag(\mu_3\text{-PTA=S})]_n(NO_3)_n \cdot nH_2O$ was the result of the combination of its constituents.

Tamames-Tabar *et al.* [49] synthesized biocompatible and bioactive Zn azelate, BioMIL-5, which inhibited the growth of *S. aureus* and *S. epidermis* due to its individual components. BioMIL-5 progressively degraded in the culture broth and water, thus, releasing active constituents. Antibacterial activity studies included the determination of MIC and MBC of Zn, azelaic acid and BioMIL-5 (MIC: 0.5, 1.5 and 1.7 ppm; MBC: 2.0, 3.0 and 4.3 ppm), where no synergic but additive activity was found, along with the determination of the duration of the effect under these concentrations. While smaller concentration (0.5 and 1.7 ppm) significantly decreased the growth rate, the inhibitory effect over bacteria was prolonged up to 7 days at the MBC concentration of 4.3 ppm. Other bioMOF, [Zn(μ-4-hzba)$_2$]$_2$·4(H$_2$O), based on Zn and hydrazinebenzoate ligands was reported by Restrepo *et al.* [54], along with the assessment of its effective antimicrobial efficiency against *S. aureus*. Diffusion of the powder on the agar plate was observed to produce higher inhibition zones than the ligand alone, whereas both exhibited similar metabolic bacterial activity inhibition, measured through the fluorescein diacetate (FDA) fluorescence staining method. Stability studies indicated that [Zn(μ-4-hzba)$_2$]$_2$·4(H$_2$O) provided a slower and more controlled release of the ligand, indicating a long-term potential. Although not experimentally proven, the combined antimicrobial activity of a Zn-based MOF, (Zn$_2$(ppa)$_2$(1,3-bdc)(H$_2$O)), coordinating to a quinolone-like drug, was also suggested in another study by Duan *et al.* [55].

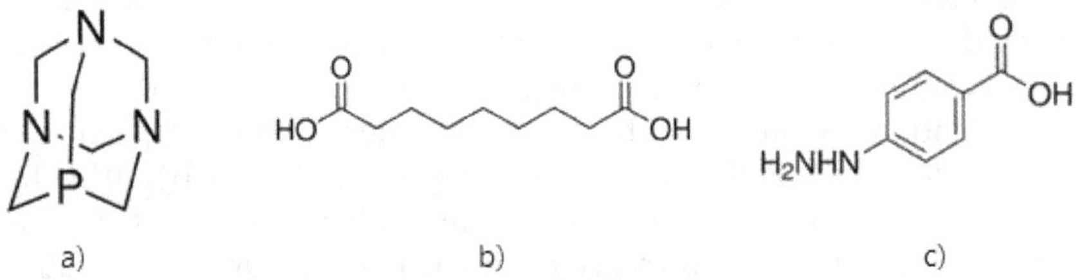

a)   b)   c)

Figure 1.5 Chemical structure of some antimicrobial ligands used in the preparation of bioMOFs: a) 1,3,5-triaza-7-phosphaadamantane (used in [53]), b) azelaic acid (used in [49]) and c) 4-hydrazinobenzoic acid (used in [54]).

Finally, the antibiotic effect of few MOFs based on potentially antimicrobial ligands has been suggested in some literature studies, however, no experimental microbiocidal results have been presented. Bio-MIL-3 [56], a porous calcium 3,3',5,5'-azobenzenetetracarboxylate with accessible metal acid sites, is able to adsorb and deliver significant amounts of NO. As Ca has *a priori* no effect on bacteria, it is suggested that a potential antimicrobial activity could result from the ligand (azobenzene derivative) [57], being further increased by the encapsulation of NO on the metal sites (see Section: Association of Therapeutic Agents). Similarly, a Ba(II) azodibenzoate, Ba(4,4'-ADB), was also reported as a potential antimicrobial MOF based on the bactericidal activity of its ligand [58].

1.3 Extrinsic Antimicrobial Activity

Although the permanent porosity of MOFs has been traditionally exploited for the capture and storage of several fluids (e.g. CO_2, CH_4, H_2, etc.) [59], an interest in the encapsulation of biologically active species (AS: biomolecules, enzymes, metallic nanoparticles, etc.) has also risen. Apart from being allocated within the framework (defects, porosity, etc.), these AS could interact with different functional groups in the ligands (e.g. $-NH_2$, $-OH$, $-SO_3H$, $-Br$, among others) *via* van der Waals forces and/or hydrophobic and electrostatic interactions [60]. In this manner, MOFs can be used as stable carriers of AS, exhibiting an extrinsic activity associated with their cargo.

In the case of MOFs used as antibacterial agents, both metallic nanoparticles and therapeutic agents have been immobilized on their structure to achieve extrinsic antimicrobial activity. Even more, when included in a MOF with intrinsic bactericidal effect, their individual actions can be complemented. The mechanism of action might result from the performance of the composite entity (AS@MOF), while on other occasions, the AS is required to be released to come in contact with the target bacteria, *via* diffusion from the MOF porosity, detachment from its surface or release subsequent to MOF degradation. In literature, the recent efforts seem to move the focus towards the development of MOF composites with extrinsic antimicrobial activity.

1.3.1 Strategies for Immobilization of Active Species

MOFs are commonly proposed as immobilizing supports due to their high surface area, robustness and compositional versatility. Three different synthetic procedures for the immobilization of AS on MOFs can be distinguished: i) diffusion, ii) *in-situ* or *de novo* synthesis and iii) chemical and physical forces [61].

The diffusion protocol, also known as 'ship-inside-a-bottle', is normally achieved in two steps: 1) MOF synthesis and activation, to ensure a free porosity and 2) post-synthetic encapsulation of the AS by diffusion through the porosity (see Figure 1.6(a)). Some limitations of this procedure are: i) the size of the AS to be encapsulated is limited by the pore dimensions, ii) the potential leaching of the encapsulated AS and iii) the precise control on the final location of AS within the framework is challenging to achieve, as a fraction might be interacting with the surface instead of being allocated in the MOF porosity. These deficiencies can be overcome with the *in-situ* synthetic protocol, as the AS is placed in contact with the MOF precursors (metal and ligand), which serves as a seed for the crystal growth of the framework surrounding them ('bottle-around-a-ship', see Figure 1.6(b)). In this case, the main constraint is the compatibility between the integrity of the AS and the MOF synthetic conditions, in order to preserve their nature and activity.

The third synthetic procedure involves the adsorption of AS on the outer MOF surface *via* different forces, either weak (van der Waals forces and/or hydrophobic and electrostatic interactions) or covalent interactions (see Figure 1.6(c)). As in the diffusion procedure, this strategy involves the use of preformed MOFs *prior* to the post-synthetic surface modification. Considering the high internal/external surface ratio, it is expected that the resulting composite has a lower AS association compared to the other procedures.

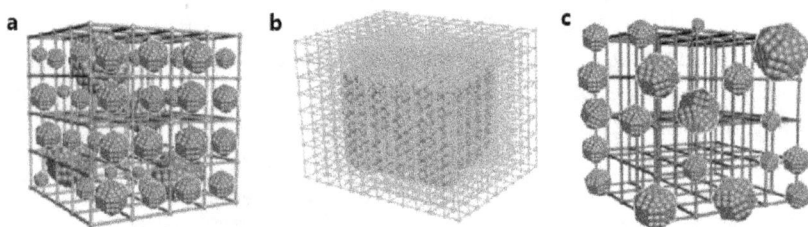

Figure 1.6 Representative structures of the AS@MOF composites achieved using different synthetic approaches. Reproduced from Reference 62 with permission from American Chemical Society.

1.3.2 Association of Metallic Nanoparticles

Noble-metal-nanoparticles@MOF core shell heterostructures have become promising candidates for many relevant applications (e.g. catalysis, imaging and sensing) [63] (Figure 1.7). In the case of antimicrobial MOF composites, Ag nano-species have been widely selected for their extensive biocidal activity [64] to achieve extrinsic bactericidal properties.

Ximing *et al.* [65] used *in-situ* synthesis to generate spherical copper porphyrin derivative MOF, CuTPP, as outer shell for previously prepared AgNPs and provided structural characterization of the composite material. Ag-CuTCPP ion release, determined *via* ICP-OES, indicated that CuTCPP MOF was stable, and the Ag composite exhibited a steady Ag+ release. *In vitro* antibacterial tests were carried out against *E. coli*, *S. aureus* and *B. subtilis*. MIC of Ag-CuTCPP against *E. coli* was higher than the control, indicating an inefficient treatment. Nevertheless, MIC for *S. aureus* and *B. subtilis* strains were within the same range as penicillin (even though the MIC values were higher than the Ag⁰ and Ag+ positive controls). After contact with Ag-CuTCPP, the SEM micrographs of planktonic bacteria suggested that the released Ag+ caused cell disruption and bacterial death by intracellular content leakage. In addition to the bactericidal properties, the potential of Ag-CuTCPP for wound healing was also investigated. For this purpose, cytotoxicity was also determined using a half inhibitory concentration (IC50) value of 50.3 µg·mL⁻¹. The material exhibited lower cytotoxicity than free AgNPs (6.5 µg·mL⁻¹). *In vivo* wound healing for infection in mice was then performed, which resulted in higher antibacterial effect of Ag-CuTCPP MOF as compared to penicillin, along with better sustenance of the activity with time (as demonstrated by colony count in heart, lung, liver and kidney).

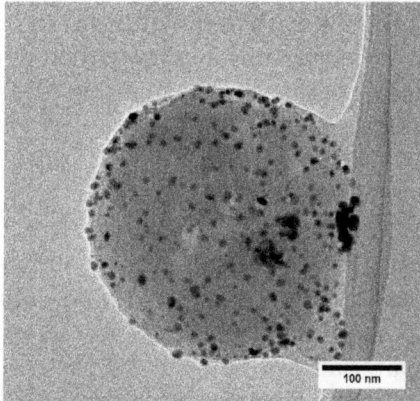

Figure 1.7 Example of a TEM micrograph of a noble-metal-nanoparticles@MOF, with the loaded NPs in higher contrast (scale bar 100 nm).

Thakare and Ramteke [66] also anchored AgNPs on MOF-5 and combined the photo- and bioactive properties against *E. coli*, inactivating 90% of bacteria in 1 h. However, the experimental conditions used for the synthesis of the material (silver content is not specified) and for the biocide experiments (*E. coli* strain, MOF concentration, irradiation λ and W) need to be described in detail.

Mortada *et al.* [67] took advantage of the free pyridyl and dicarboxyl groups present in the Zr-based MOFs UiO-67-bpydc and UiO-66-2COOH (using 2,2'-bipyridine-5,5'-dicarboxylate and 1,2,4,5-benzenetetracarboxylate as ligands) respectively, to metalate silver cations by post-synthetic modification. Metalation was confirmed by a combination of experimental techniques, determining an Ag content of 12 and 13 wt% in UiO-67-bpydc and UiO-66-2COOH respectively. OD measurements enabled the determination of MIC, which was observed to be 50 ppm for UiO-67-bpydc and 75 ppm for UiO-66-2COOH, corresponding to 6.5 and 9.6 ppm of Ag respectively. These results are observed to be comparable to the Ag-based MOFs synthesized by Lu *et al.* [29].

Guo *et al.* [68] employed an *in-situ* procedure to prepare a core-shell composite consisting of pre-synthesized Ag nanowires with Zn-imidazolate ZIF-8 and tested antibacterial activity against *B. subtilis* and *E. coli*. MIC, determined by OD_{600}, indicated that 200 and 300 ppm of Ag@ZIF-8, corresponding to 80 and 120 ppm of Ag respectively, were required to prevent the growth of bacterial colonies of *B. subtilis* and *E. coli* respectively. Positive controls of Ag nanowires and pristine ZIF-8 generated with the same MIC concentrations exhibited lower antibacterial efficacy than Ag@ZIF-8, suggesting a synergistic effect of the Ag core and the ZIF-8 shell, which stabilized the silver release.

1.3.3 Association of Therapeutic Agents

Studies on the cytotoxicity of MOFs [69] have indicated the potential of using MOFs as therapeutic carriers, with potential applications in drug delivery and as antimicrobial agents. Even more, bioMOFs, used as hosting material, can be used for the concomitant administration of more than one therapeutic agent.

Early work in this field considered the encapsulation of NO (an important biological signaling molecule), owing to its antibacterial and wound healing activities [70]. NO adsorption and release was thoroughly studied using MOFs, in particular, metal hydroxyterephthalate M-CPO-27 (M= Co, Ni, Zn; also called MOF-74) and HKUST-1 [71]. McKinlay *et al.* [72] presented the activity of Ni-CPO-27, loaded simultaneously with NO and the antibiotic metronidazole, against *P. aeruginosa* and *S. aureus*. The multi-rate delivery of the bactericidal agents was assessed, with NO release faster than the larger metronidazole molecules. Antibacterial activity tests revealed the relatively effective intrinsic biocidal activity of the pristine MOFs. Encapsulation of metronidazole solely resulted in slight reduction of the intrinsic activity due to the reduced ability to release metal ions, as the channels were blocked with the antibiotic. Nevertheless, the inclusion of NO and metronidazole significantly increased the inhibition of metabolic activity, thus, resulting in bactericidal performance even at shorter durations due to the fast NO release, along with sustenance of the material activity up to 10 days.

Iodine, another known bioactive molecule [73], was successfully encapsulated in ZIF-8 by Au-Duong and Lee [74] to benefit from the MOF degradation under acidic pH for progressive release. Both iodine adsorption and release at pH=6 were studied prior to the biocide tests. Firstly, the activity of iodine and I@ZIF-8 was demonstrated by inhibition zone against *E. coli*, *S. epidermis* and *S. aureus*, and the composite was active in plates with pH preadjusted to 6 (MOF dissolution), but not pH=7. Bacterial viability and growth inhibition, determined by colony counting, showed that the three strains were neutralized with a dosage of 200 ppm of I@ZIF-8 after 3 minutes only. Further, I@ZIF-8 thin films deposited over glass slides were used to determine the inhibition of biofilm formation. After dropping 100 µL of *E. coli* and *S. aureus* bacterial suspensions at pH 6 and 7, slides were observed with only LIVE/DEAD stain under confocal microscope. Images revealed the loss of membrane integrity (red stain) at pH=6, indicating again the pH-controlled iodine liberation.

The use of ZIF-8 for drug encapsulation has also been explored in other studies. Nabipour *et al.* [75] made use of the pH-dependent stability of ZIF-8 to release the antibiotic ciprofloxacin (CIP), previously encapsulated by diffusion. CIP release in pH 7.4 and 5.0 buffers was monitored by UV-vis spectroscopy and the antimicrobial activity against *E. coli* and *S. aureus* was determined by disk diffusion as well as determination of the inhibition zone. At pH 5, almost 70% of the CIP was released within the first 5 h, while at pH 7.4, as ZIF-8 was more stable, 70% release of CIP took 48 h. The larger inhibition zone of CIP-ZIF-8, compared to the free CIP and the pristine MOF, was associated with the gradual degradation of ZIF-8 and the release of Zn^{2+} and CIP. Chowdhuri *et al.* [76] co-encapsulated in one-pot synthesis procedure vancomycin (VAN) and folic acid (FA) in ZIF-8 to treat multi-drug resistant *S. aureus* and *E. coli*. MBC of ZIF-8@FA@VAN against *S. aureus* was 16 ppm and >512 ppm for *E. coli*. With the help of the conjugated carbon dots and the evaluation of their fluorescent emission by confocal fluorescence microscopy, it was observed that ZIF-8@FA@VAN internalization was significantly higher in the case of *S. aureus*, explaining the greater biocide activity of the composite against this strain. In addition, *S. aureus* bacteria treated with ZIF-8@FA@VAN exhibited a higher ROS generation, indicating it to be the possible mechanism of cell death. SEM images of treated *S. aureus* revealed a damaged surface compared to the control, while no significant change was observed in *E. coli*. These results suggested the potential use of ZIF-8@FA@VAN for the treatment of challenging multi-drug resistant *S. aureus*.

Nabipour and coworkers proposed another Zn-based MOF for the encapsulation of different drugs with antibacterial properties. The authors explored the post-synthetic inclusion of nalidixic acid in the porosity of $Zn_2(bdc)_2(dabco)$ [77], where the Zn cluster was linked to bencenedicarboxylates and diazabicyclooctanate. The activity of the developed MOF against *E. coli* and *S. aureus* was analyzed. Inhibition zone test indicated a higher inhibition in the case of *E. coli*, higher than the inhibition of the pristine MOF and drug alone in both strains. Antibacterial activity revealed a MIC value lower than 0.05 ppm for both strains. More recently, the authors also presented $Zn_2(bdc)_2(dabco)$, loaded with gentamicin [78], immobilized by diffusion. In this case, only the inhibition zone was evaluated, being higher for *S. aureus*. From the results of both studies, it is reasonable to derive the importance of the selection of the drug to be immobilized for attributing extrinsic antimicrobial activity to a MOF composite.

Claes *et al.* [79] explored the stimulus-responsive degradation of the porous iron(III) terephthalate MIL-88B(Fe) to release antibiofilm compounds (Figure 1.8). *Salmonella enterica* subspecies *S. Typhimurium* secretes siderophores that have a high affinity for iron and can produce a controlled release of the inhibitor encapsulated in the MOF lattice by metal complexation. 5-(4-Chlorophenyl)-*N*-(2-isobutyl)-2-aminoimidazole (IMI) was the selected antibiofilm compound, with a 10 wt% encapsulated by diffusion in a previously activated MIL-88B(Fe) solid. MOF degradation in water and in presence of an external chelator (Na_3 citrate) was verified. The results showed that the presence of IMI within the porosity did not affect the stimulus-responsive degradation and that IMI was easily trigger released. To test the inhibition of biofilm, the MIL-88B(Fe) micrometric particles were deposited by drop-casting on polystyrene petri-dishes. Non-loaded MIL-88B(Fe) is non-toxic to biofilm. With 1 mg·cm^{-2} concentration of MOF loaded with 10 wt% of IMI, 70±10 % of *S. Typhimurium* were inhibited without the need of external chelators, compared with the non-coated petri dish. Interestingly, by increasing the MOF coating to 1.25 mg·cm^{-2}, almost no biofilm was grown on the MOF surface.

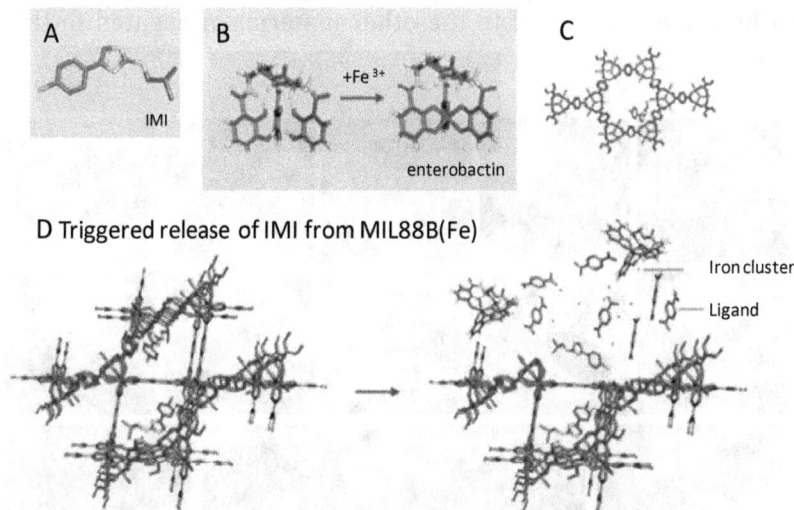

Figure 1.8 Anti-biofilm compound release from MIL-88B(Fe) by competitive metal complexation. Depicted are (A) the chemical structure of the anti-biofilm compound (IMI), (B) chelation of Fe(III) by the siderophore enterobactin, (C) adsorption of IMI in MIL-88B(Fe) visualized from the c-axes and (D) the triggered release of IMI from MIL-88B(Fe). Hydrogen atoms are hided except for the enterobactin molecules in (B) and (D) (C = dark grey; O = red; N = green; Fe = purple; H = light grey; Cl = orange). Reproduced from Reference 79 with permission from American Chemical Society.

Also made from iron, the microporous biocompatible carboxylate MOF-53 NPs were used by Lin *et al.* [80] as platform to encapsulate VAN by physisorption and release it due to the pH-dependent MOF degradation (Figure 1.9). Release studies in PBS revealed that more than 90% of VAN was released at pH 7.4 and 6.5 after 24 h, while at pH 5.5, around 75% VAN was released during the same period. Bacterial inhibition of *S. aureus* suspensions against VAN@MOF-53 with 20 wt% of VAN was determined by CFU plate count, determining that bacteria were no longer viable with 200 ppm of VAN@MOF-53. In addition, the absence of cytotoxicity of VAN@MOF-53 was determined using an MTT assay carried with calvarial cells of mice. Nase-ri *et al.* [81] presented a new Cu-based bioMOF with 4,4'-biphenyldicarboxylate as ligand and cytosine as extrinsic antimicro-bial. Cytosine was embedded with a simple one-step sonochemical method. MIC and MBC were determined against *Proteus mirabilis* (1600 and 2000 ppm respectively), which were observed to be rela-tively higher as compared to the other materials presented in this review.

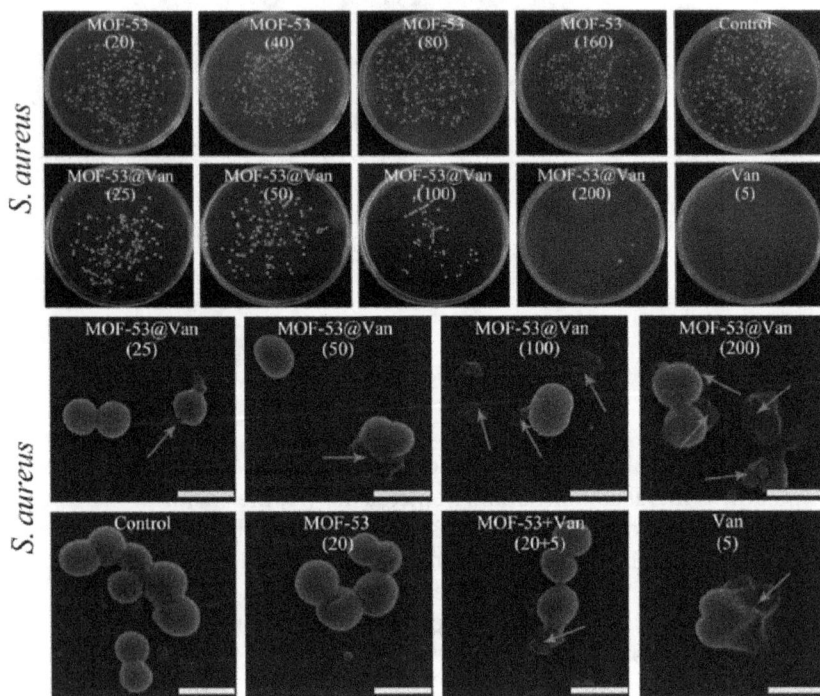

Figure 1.9 (Top): Representative images of viable S. aureus grown on different samples after 24 h of culture ; (Bottom): SEM morphology of S. aureus seeded on various samples after incubation at 37 °C for 24 h. The scale bar is 1 μm. Reproduced from Reference 80 with permission from American Chemical Society.

1.4 Devices and Applications

In most of the literature presented so far, MOFs were handled as powder suspensions in contact with the target bacteria (in the culture broth suspension or in the agar plate). Nevertheless, polycrystalline powders are unsuitable for industrial applications due to difficulties in processing (dust, clogs, mass loss, challenges during transfer, etc.) [82]. Consequently, it is of great interest to shape MOFs into more application-oriented forms such as pellets and films, along with their inclusion into composite materials [83]. It can be achieved either during the synthesis process (such as monoliths of MOFs [84]) or by the immobilization on already shaped substrates [85]. In this respect, the MOF-shaping strategies followed to date targeting antimicrobial applications are presented in the following sections.

1.4.1 Thin Films and Surface Coatings

Thin films and surface coatings involve different techniques, including drop-casting, spin-coating, layer-by-layer deposition and crystal growth. An example is the previously mentioned Cu-SURMOF-2 [39], prepared layer-by-layer *via* spray deposition over 16-mercaptohexadecanoic acid self-assembled monolayers (MHDA-SAM) on top of gold substrates and proposed as antifouling coating against *C. marina* (Figure 1.10). Bacterial viability, determined by

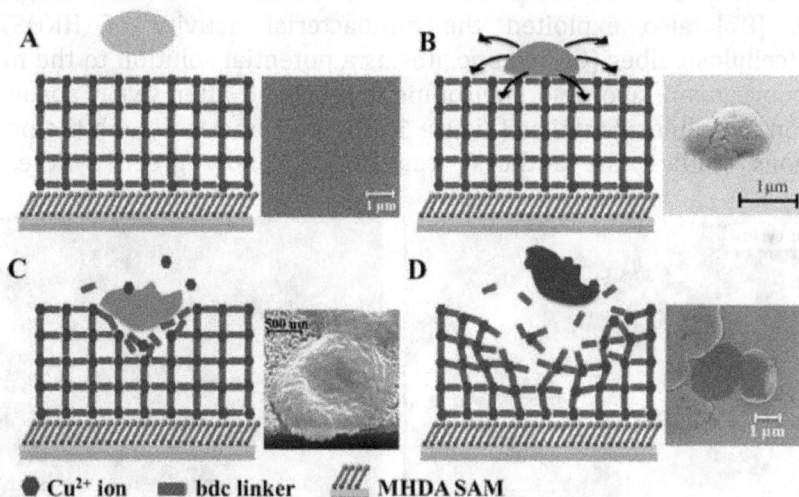

$\bullet$ Cu^{2+} ion   ■ bdc linker   ⁄⁄⁄ MHDA SAM

Figure 1.10 Schematic illustration of the active disassembly of the Cu-SURMOF 2 under the influence of *C. marina* and SEM images of the bacteria in contact with the surface. Reproduced from Reference 39 with permission from Springer.

LIVE/DEAD, revealed 8% of yellow stained bacteria, commonly accepted as debilitated bacteria, but still alive. In addition, the authors determined that a Cu-SURMOF coated surface reduced by 50% the necessary shear stress required to remove bacteria biofilm, as compared with a MHDA-SAM positive blank. Other MOFs studied as surface coatings for biofilm inhibition are I@ZIF-8 [74] and IMI@MIL-88B [79], with thin films prepared by simple drop-casting.

In order to obtain antimicrobial textiles for wound healing and surgical procedures, Abbasi *et al.* [35] reported an ultrasonically-assisted synthesis of HKUST-1 as surface coating of silk fibers *via* several dipping cycles. Antimicrobial tests were performed using the inhibition zone test against *E. coli* and *S. aureus*. While the pristine silk fibers showed no inhibition, moderate antibacterial activity was obtained in the composite due to the slow Cu ion release. Rodriguez *et al.* [86] also explored HKUST-1 as antimicrobial surface coating using previously electrospun cellulosic fibers pretreated with carboxylic groups and a layer-by-layer alternative dipping protocol. A preliminary inhibition zone test against *E. coli* indicated that the bacterial inhibition resulted from the Cu ions released from the MOF coating. CFU plate count of *E. coli* bacterial suspensions incubated for 1 h in contact with 2x2 cm^2 cellulosic patches coated with HKUST-1 revealed complete bacterial viability inhibition. Wang *et al.* [87] also exploited the antibacterial activity of HKUST-1/cellulosic fiber (CF) composites as a potential solution to the microorganism favorable environment produced after sweat absorption in cellulosic textiles (Figure 1.11). In comparison with the previous studies, the authors presented a totally green synthesis

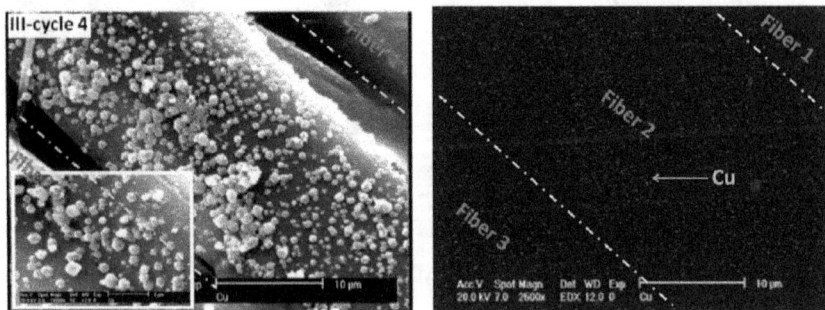

Figure 1.11 SEM image and wavelength-dispersive X-ray (WDX) analysis of group III fibers after 4 cycles of dipping. Reproduced from Reference 35 with permission from Elsevier.

procedure which avoided the use of DMF as solvent. Antibacterial activity was confirmed *via* disk diffusion and inhibition of *E. coli* and *S. aureus*.

Another Cu-based MOF reported as antifouling coating is the water stable copper triazolate Cu-BTTri [88], dispersed in a chitosan film (10 wt% of MOF). The authors explored the combination of the antibacterial action of chitosan and Cu present in the MOF against *P. aureginosa* biofilm. Cell viability of the attached bacteria was determined after 6 and 24 h by means of a spectroscopic assay called CTB, which is an indirect measurement of the metabolic activity of bacteria. While chitosan films inhibited ~55% of bacteria, the addition of Cu-BTTri composite film reached ~85% inhibition, with further sustenance over 24 h period. Reusability test of the films indicated no statistical difference between the first and second cycles. After ligand and copper release studies, it was concluded that the activity of the Cu-BTTri/chitosan film resulted from the constituent leaching. In spite of the interesting bactericidal studies on sessile *P. aureginosa* bacteria in the biofilm, the antifouling properties of the Cu-BTTri/chitosan film were not reported.

Similarly, Co-based MOFs have been mixed with fibers to provide them antimicrobial properties. Recently, Qian *et al.* [89] presented cellulose paper composite with ZIF-67 (CP/CNF/ZIF-67) as a biodegradable biocidal paper for packaging, generated by following an impregnation-crystal growth protocol. The mechanical properties of CP/CNF/ZIF-67 composite evidenced an improved tensile strength, elastic modulus, folding endurance and tear index compared to pristine fibers, thus, suggesting that the MOF NPs acted as crosslinkers, binding to the surrounding cellulose fibers. Antibacterial tests of CP/CNF/ZIF-67 against *E. coli* indicated a promising inhibition zone that increased with the number of ZIF-67 NPs on the fibers' surface. In the study of Quirós *et al.* [46], direct blending of Co-SIM-1 with PLA was employed to produce electrospun mats made of composite fibers, where the MOF microscale particles were completely embedded inside the polymeric fibers. The antimicrobial potential of Co-SIM-1/PLA mats was evaluated against *P. putida* and *S. aureus* biofilm formed on the mat surface. SEM images revealed the biofilm inhibition (up to 30% with composites based on 6 wt% content of Co-SIM-1) and LIVE/DEAD staining confirmed bacterial viability inhibition, being more apparent in the case of *P. putida*. Another imidazolate MOF, ZIF-8, was presented by Miao *et al.* [90] in a mixture with polyvinylidenefluoride (PVDF) and Hperfluorooctyltriethoxysilane (POTS) to obtain a superhydrophobic self-cleaning and antibacterial coating for different substrates (e.g. membranes, filter paper, non-woven fabrics and textiles). Bacterial inhibition of *E. coli* suspensions in contact with 10 cm^2 of coated substrates, determined by CFU plate count, indicated that the Zn ion release might be the cause of the important biocidal activity of the composite coating.

1.4.2 Membranes

MOF thin film processing in the active layer of microfluid membranes has also been extensively studied. Generally, polymeric or inorganic hollowed or fibred substrates are selected and the MOF particles are grown by solvo/hydrothermal synthesis (*in-situ* or seeded growth), interfacial synthesis in two immiscible solvents and liquid phase epitaxy (layer-by-layer) method [91] (Figure 1.12).

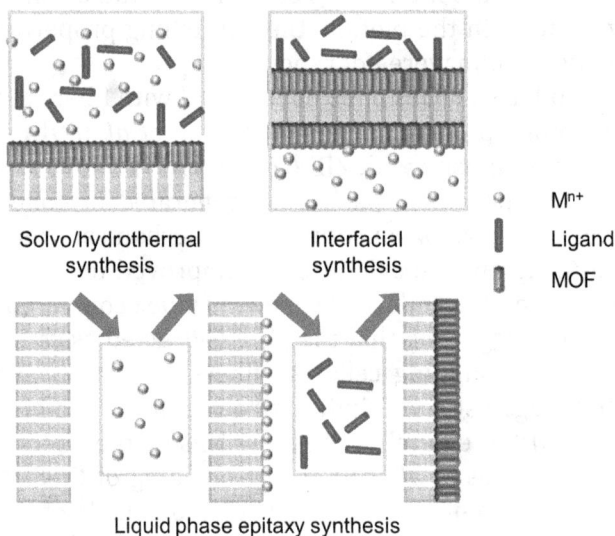

Figure 1.12 Scheme of the synthesis methods for continuous MOF-composite membranes.

Zirehpour *et al.* [50] incorporated silver trimesate MOF nanoparticles (Ag-BTC, 33 nm) in the active layer of a forward osmosis membrane (0.02%) to mitigate biofouling [50]. Characterization of the composite exhibited a good compatibility of the MOF with the active thin-film polyamide matrix of the thin film nanocomposite

(TFN) membrane. Bactericidal activity of the Ag-BTC/TFN membranes was studied against *S. aureus* and *E. coli* biofilms. Bacterial viability was determined from the plate count of the colony forming units of the bacteria detached from the surface after 1 h incubation of the inoculums in contact with the membrane active layer. Results showed a 90% bacterial viability inhibition for both strains with a "fresh" Ag-BTC/TFN membrane as well as with the membrane pre-soaked in water for 24 h and 6 months, suggesting a remarkable long-term bactericidal activity. It is important to notice that only 1 h contact might not be representative for sessile bacterial inhibition, as longer time is usually required to develop a mature biofilm [92]. In addition, Ag+ release studies indicated a burst release of silver within the first day that might be responsible for the biocidal effect. Biofouling experiments were determined by passing an *E. coli* solution and analyzing the water flux through the membrane during 24 h. While the flux of the TFN membrane (negative blank) decreased by 20% after 24 h, the Ag-BTC/TFN membrane reduced the flux by 10% only, thus, supporting the antifouling activity of the composite.

Wang *et al.* [93] prepared as well the TFN membranes *via* simple interfacial polymerization, integrating both ZIF-8 and graphene oxide in the active layer (Figure 1.13). The authors demonstrated the antimicrobial properties of the composite membranes (TFN-ZG) against *E. coli*. The presence of graphene oxide prevented the agglomeration of MOF NPs in the active layer. Determination of the inhibition concentration for inoculums in suspension (by plate count) showed the lower bacterial viability when the content of ZIF-8 increased (TFN-ZG1: 50%, TFN-ZG2: 66%, TFN-ZG3: 84% bacterial inhibition). Besides, hydrophilicity, determined by contact angle, had a significant influence over the antifouling and filtration performance of the TFN-ZG membranes. TFN-ZG1 exhibited smaller contact angle (10.7°), and the hydrophobic nature of ZIF-8 prevailed with higher amounts of MOF.

Finally, Ming *et al.* [94] prepared a MOF-polymer mixed matrix membrane (MMMs) based on poly(ε-caprolactone) (PCL) and the Zr porphyrin-MOF MOF-525. This MOF was selected due to its potential to generate ROS after irradiation. The MOF-525/PCL membrane was tested as photodynamic bactericidal agent against *E. coli*. To ensure good compatibility between the MOF particles and polymeric matrix, 10 or 30 wt% of 100 nm MOF NPs (synthesized by optimized solvothermal procedures) were used. The resulting MOF-525/PCL membrane possessed a homogenous red color, characteristic of porphyrin derivatives. The homogeneous distribution of the MOF NPs was confirmed by SEM and confocal microscopic studies. In addition, a good biocompatibility of the MOF-525/PCL membrane without irradiation was confirmed by plate count. Nevertheless, upon irradiation, a noticeable inhibition effect was evidenced within the first 30 min, increasing with the MOF-525 content in the membrane (reaching even 50% of the bacterial inhibition with 30 wt% MOF content, as determined by the plate count).

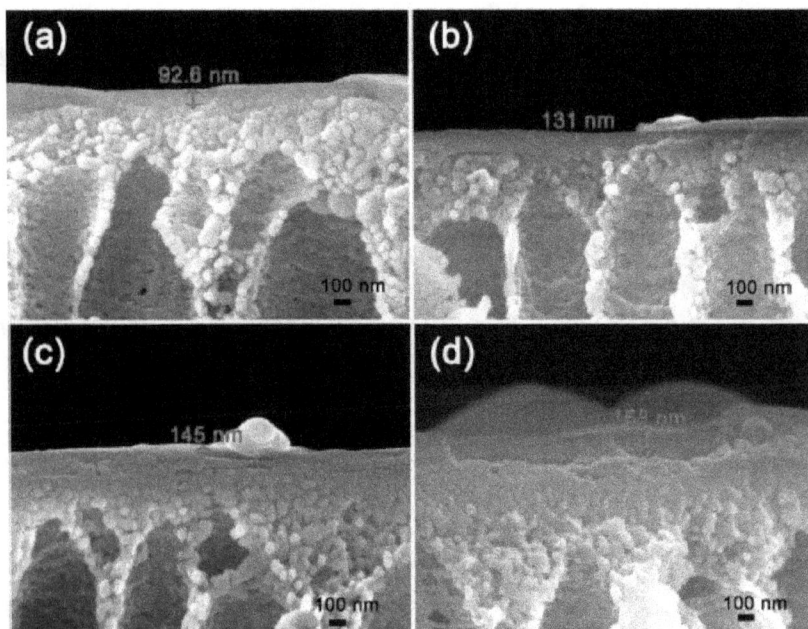

Figure 1.13 FE-SEM images of the cross section of the membranes: pristine (a), TFN-ZG1 (b), TFN-ZG2 (c), and TFN-ZG3 (d). Reproduced from Reference 93 with permission from American Chemical Society.

1.5 Conclusions and Perspectives

MOFs, due to their compositional versatility and porosity, have generated a significant interest as antimicrobial agents. Early attempts studied their intrinsic biocidal activity, by the selection of bioactive cations and/or organic ligands. Due to the extensively known activity of silver, Ag-based MOFs have been widely studied, evidencing their bactericidal properties against both Gram (+) and Gram (-)

bacteria. Other frequently proposed MOFs are the microporous copper trimesate HKUST-1, used as bactericidal and antifungal agents as well as the zinc-imidazolate ZIF-8, easily synthesized at room temperature. Though studied to a lesser degree, some cobalt imidazolates (ZIF-67, Co-SIM-1, etc.) also seem to be promising biocidal materials. Further, the combination of an active cation with a bioactive organic species leads to concurrent bactericidal MOFs, such as the Zn azelate bioMIL-5. In spite of the interest in this multi-target approach, this research direction is at an early stage, and there are plenty opportunities for the development of new MOF structures with combined (additive or even synergistic) antimicrobial activity. Important selection criteria might be the use of endogenous molecules to overcome toxicity when the MOF is degraded to release the bactericidal compounds to the media, particularly in *in vivo* applications.

However, there has been a change of trend towards the integration of bioactive species (e.g. metallic nanoparticles, therapeutic agents, etc.) within the MOF porosity to provide extrinsic antimicrobial activity. Even more, when included in a MOF with intrinsic bactericidal effect, their individual actions can be complemented. Up to now, only metallic silver, in the form of nanoparticles or nanowires, has been employed in different studies. In contrast, a large variety of antimicrobial molecules have been entrapped within MOF porosity (e.g. NO, I, ciprofloxacin and vancomycin). It is envisaged that the low cytotoxicity of some MOFs will enable their use as efficient drug carriers in the near future.

Both intrinsic and extrinsic biocidal MOFs need to be integrated in different application-oriented forms for their final use. Common solutions are the elaboration of thin films or surface coatings and the MOF integration in substrates such as membranes. A good compatibility between the MOF and substrate is the key for the final device integrity. These devices are the ones ultimately in contact with the bacterial environment and should preserve their antimicrobial properties over time to ensure their action. The manufacturing of these devices has lately enabled the study of MOF bactericidal activity not only against bacterial suspensions, but also against sessile bacteria in the challenging biofilms, which exhibit higher resistance against environmental stresses and traditional antibiotics. Therefore, progress in MOF conformation will also guide the successful attainment of the relevant purpose of the bacteria-free surfaces.

There is still a significant leap needed from the lab scale till final biocidal devices. The studies so far have proven the potential of MOFs as effective antimicrobial agents, which needs to be translated to actual devices. Besides biomedical applications (e.g. surgical implants, wound healing, etc.), biocidal MOFs are also promising for water treatment, food packaging, maritime transport and off-shore industries, heat exchangers, electrochemical devices, among others.

References

1. Even, C., Marlière, C., Ghigo, J.-M., Allain, J.-M., Marcellan, A., and Raspaud, E. (2017) Recent advances in studying single bacteria and biofilm mechanics. *Advances in Colloid and Interface Science*, **247**, 573-588.
2. Liu, K., and Jiang, L. (2012) Bio-inspired self-cleaning surfaces. *Annual Review of Materials Research*, **42**(1), 231-263.
3. Okeke, I. N., Lamikanra, A., and Edelman, R. (1999) Socioeconomic and behavioral factors leading to acquired bacterial resistance to antibiotics in developing countries. *Emerging Infectious Diseases*, **5**(1), 18-27.
4. Townsley, L., and Shank, E. A. (2017) Natural-product antibiotics: cues for modulating bacterial biofilm formation. *Trends in Microbiology*, **25**(12), 1016-1026.
5. Stewart, P. S., and William Costerton, J. (2001) Antibiotic resistance of bacteria in biofilms. *Lancet*, **358**(9276), 135-138.
6. Bordi, C., and de Bentzmann, S. (2011) Hacking into bacterial biofilms: a new therapeutic challenge. *Annals of Intensive Care*, **1**(1), 19.
7. Simchi, A., Tamjid, E., Pishbin, F., and Boccaccini, A. R. (2011) Recent progress in inorganic and composite coatings with bactericidal capability for orthopaedic applications. *Nanomedicine: Nanotechnology, Biology and Medicine*, **7**(1), 22-39.
8. Raquez, J.-M., Habibi, Y., Murariu, M., and Dubois, P. (2013) Polylactide (PLA)-based nanocomposites. *Progress in Polymer Science*, **38**(10-11), 1504-1542.
9. Wang, Z., Shen, Y., and Haapasalo, M. (2014) Dental materials with antibiofilm properties. *Dental Materials*, **30**(2), e1-e16.
10. Férey, G. (2008) Hybrid porous solids: past, present, future. *Chemical Society Reviews*, **37**(1), 191-214.
11. Farha, O. K., Eryazici, I., Jeong, N. C., Hauser, B. G., Wilmer, C. E., Sarjeant, A. A., Snurr, R. Q., Nguyen, S. T., Yazaydın, A. Ö., and Hupp, J. T. (2012) Metal-organic framework materials with ultrahigh surface areas: is the sky the limit? *Journal of the American Chemical*

Society, **134**(36), 15016-15021.

12. Furukawa, H., Cordova, K. E., O'Keeffe, M., and Yaghi, O. M. (2013) The chemistry and applications of metal-organic frameworks. *Science*, **341**(6149), 1230444.

13. Horcajada, P., Gref, R., Baati, T., Allan, P. K., Maurin, G., and Couvreur, P. (2012) Metal - organic frameworks in biomedicine. *Chemical Reviews*, **112**, 1232-1268.

14. Wyszogrodzka, G., Marszałek, B., Gil, B., and Dorożyński, P. (2016) Metal-organic frameworks: mechanisms of antibacterial action and potential applications. *Drug Discovery Today*, **21**(6), 1009-1018.

15. Doonan, C., Riccò, R., Liang, K., Bradshaw, D., and Falcaro, P. (2017) Metal-organic frameworks at the biointerface: synthetic strategies and applications. *Accounts of Chemical Research*, **50**(6), 1423-1432.

16. Berchel, M., Gall, T. Le, Denis, C., Hir, S. Le, Quentel, F., Elléouet, C., Montier, T., Rueff, J.-M., Salaün, J.-Y., Haelters, J.-P., Hix, G. B., Lehn, P., and Jaffrès, P.-A. (2011) A silver-based metal-organic framework material as a 'reservoir' of bactericidal metal ions. *New Journal of Chemistry*, **35**(5), 1000-1003.

17. Rojas, S., Devic, T., and Horcajada, P. (2017) Metal organic frameworks based on bioactive components. *Journal of Materials Chemistry B*, **5**(14), 2560-2573.

18. Slenters, T. V., Sagué, J. L., Brunetto, P. S., Zuber, S., Fleury, A., Mirolo, L., Robin. A. Y., Meuwly, M., Gordon, O., Lamdmann, R., Daniels, A. U., and Fromm, K. M. (2010) Of chains and rings: Synthetic strategies and theoretical investigations for tuning the structure of silver coordination compounds and their applications. *Materials*, **3**(5), 3407-3429.

19. Brunetto, P. S., Slenters, T. V., and Fromm, K. M. (2010) In vitro biocompatibility of new silver(I) coordination compound coated-surfaces for dental implant applications. *Materials*, **4**(2), 355-367.

20. Wang, K., Yin, Y., Li, C., Geng, Z., and Wang, Z. (2011) Facile synthesis of zinc(II)-carboxylate coordination polymer particles and their luminescent, biocompatible and antibacterial properties. *CrystEngComm*, **13**(20), 6231-6236.

21. Tăbăcaru, A., Pettinari, C., Marchetti, F., Di Nicola, C., Domasevitch, K. V., Galli, S., Masciocchi, N., Scuri, S., Grappasonni, I., and Cocchioni, M. (2012) Antibacterial action of 4,4'-bipyrazolyl-based silver(I) coordination polymers embedded in PE disks. *Inorganic Chemistry*, **51**(18), 9775-9788.

22. Chernousova, S., and Epple, M. (2013) Silver as antibacterial agent: Ion, nanoparticle, and metal. *Angewandte Chemie - International Edition*, **52**(6), 1636-1653.

23. Raghupathi, K. R., Koodali, R. T., and Manna, A. C. (2011) Size-dependent bacterial growth inhibition and mechanism of antibac-

terial activity of zinc oxide nanoparticles. *Langmuir*, **27**(7), 4020-4028.

24. Hong, R., Kang, T. Y., Michels, C. A., and Gadura, N. (2012) Membrane lipid peroxidation in copper alloy-mediated contact killing of Escherichia coli. *Applied and Environmental Microbiology*, **78**(6), 1776-1784.

25. Giménez-Marqués, M., Hidalgo, T., Serre, C., and Horcajada, P. (2016) Nanostructured metal-organic frameworks and their bio-related applications. *Coordination Chemistry Reviews*, **307**, 342-360.

26. Liu, Y., Xu, X., Xia, Q., Yuan, G., He, Q., and Cui, Y. (2010) Multiple topological isomerism of three-connected networks in silver-based metal-organoboron frameworks. *Chemical Communications*, **46**(15), 2608-2610.

27. Singleton, R., Bye, J., Dyson, J., Baker, G., Ranson, R. M., and Hix, G. B. (2010) Tailoring the photoluminescence properties of transition metal phosphonates. *Dalton Transactions*, **39**(26), 6024-6030.

28. Lu, X., Ye, J., Sun, Y., Bogale, R. F., Zhao, L., Tian, P., and Ning, G. (2014) Ligand effects on the structural dimensionality and anti-bacterial activities of silver-based coordination polymers. *Dalton Transactions*, **43**(26), 10104-10113.

29. Lu, X., Ye, J., Zhang, D., Xie, R., Feyisa, R., Sun, Y., Zhao, L., Zhao, Q., and Ning, G. (2014) Silver carboxylate metal - organic frameworks with highly antibacterial activity and biocompatibility. *Journal of Inorganic Biochemistry*, **138**, 114-121.

30. Wang, X., Zhao, D., Tian, A., and Ying, J. (2014) Three 3D silver-bis(triazole) metal-organic frameworks stabilized by high-connected Wells-Dawson polyoxometallates. *Dalton Transactions*, **43**(13), 5211-5220.

31. Zhang, S.-S., Wang, X., Su, H.-F., Feng, L., Wang, Z., Ding, W.-Q., Blatov, V. A., Kurmoo, M., Tung, C.-H., Sun, D., and Zheng, L.-S. (2017) A water-stable Cl@Ag $_{14}$ cluster based metal-organic open framework for dichromate trapping and bacterial inhibition. *Inorganic Chemistry*, **56**(19), 11891-11899.

32. Raffi, M., Mehrwan, S., Bhatti, T. M., Akhter, J. I., Hameed, A., Yawar, W., and ul Hasan, M. M. (2010) Investigations into the antibacterial behavior of copper nanoparticles against Escherichia coli. *Annals of Microbiology*, **60**(1), 75-80.

33. Wang, K., Geng, Z., Yin, Y., Ma, X., and Wang, Z. (2011) Morphology effect on the luminescent property and antibacterial activity of coordination polymer particles with identical crystal structures. *CrystEngComm*, **13**, 5100-5104.

34. Chui, S. S.-Y., Lo, S. M.-F., Charmant, J. P. H., Orpen, A. G., and Williams, I. D. (1999) A chemically functionalizable nanoporous material. *Science*, **283**(5405), 1148-1150.

35. Abbasi, A. R., Akhbari, K., and Morsali, A. (2012) Dense coating of surface mounted CuBTC metal-organic framework nanostructures on silk fibers, prepared by layer-by-layer method under ultrasound irradiation with antibacterial activity. *Ultrasonics Sonochemistry*, **19**(4), 846-852.

36. Chiericatti, C., Basilico, J. C., Zapata Basilico, M. L., and Zamaro, J. M. (2012) Novel application of HKUST-1 metal-organic framework as antifungal: Biological tests and physicochemical characterizations. *Microporous and Mesoporous Materials*, **162**, 60-63.

37. Celis-Arias, V., Loera-Serna, S., Beltrán, H. I., Álvarez-Zeferino, J. C., Garrido, E., and Ruiz-Ramos, R. (2018) The fungicide effect of HKUST-1 on *Aspergillus niger* , *Fusarium solani* and *Penicillium chrysogenum*. *New Journal of Chemistry*, **42**(7), 5570-5579.

38. Azad, F. N., Ghaedi, M., Dashtian, K., Hajati, S., and Pezeshkpour, V. (2016) Ultrasonically assisted hydrothermal synthesis of activated carbon-HKUST-1-MOF hybrid for efficient simultaneous ultrasound-assisted removal of ternary organic dyes and antibacterial investigation: Taguchi optimization. *Ultrasonics Sonochemistry*, **31**, 383-393.

39. Arpa Sancet, M. P., Hanke, M., Wang, Z., Bauer, S., Azucena, C., Arslan, H. K., Heinle, M., Gliemann, H., Woll, C., and Rosenhahn, A. (2013) Surface anchored metal-organic frameworks as stimulus responsive antifouling coatings. *Biointerphases*, **8**(1), 1-35.

40. Chang, E. L., Simmers, C., and Knight, D. A. (2010) Cobalt complexes as antiviral and antibacterial agents. *Pharmaceuticals*, **3**(6), 1711-1728.

41. *Zinc Sulfate Heptahydrate Safety Data Sheet*, Fisher Scientific (2010). Online: https://www.fishersci.com/store/msds?partNumber=Z76500and productDescription=ZINC+SULFA+HEPTA+USP%2FCC+500GMandvendorId=VN 00033897andcountryCode=USandlanguage=en [accessed 6th September 2018].

42. *Cobalt Sulfate Heptahydrate Safety Data Sheet*, Acros Organics (2004). Online: https://www.ch.ntu.edu.tw/~genchem99/msds/exp18/CoSO4.pd f [accessed 6th September 2018].

43. *HERA Biocompatibility Brochure*, Kulzer GmbH (2017). Online: https://www.kulzer.com/media/webmedia_local/downloads_new /hera_11/nichtedelmetalllegierungen_fuer_k_b/Hera_Biocompatib ility_Brochure_EN.pdf [accessed 6th September 2018].

44. Zhuang, W., Yuan, D., Li, J.-R., Luo, Z., Zhou, H.-C., Bashir, S., and Liu, J. (2012) Highly potent bactericidal activity of porous metal-organic frameworks. *Advanced Healthcare Materials*, **1**(2), 225-238.

45. Aguado, S., Quirós, J., Canivet, J., Farrusseng, D., Boltes, K., and Rosal, R. (2014) Antimicrobial activity of cobalt imidazolate metal - organic frameworks. *Chemosphere*, **113**, 188-192.

46. Quirós, J., Boltes, K., Aguado, S., de Villoria, R. G., Vilatela, J. J., and Rosal, R. (2015) Antimicrobial metal-organic frameworks incorporated into electrospun fibers. *Chemical Engineering Journal*, **262**, 189-197.

47. Sirelkhatim, A., Mahmud, S., Seeni, A., Kaus, N. H. M., Ann, L. C., Bakhori, S. K. M., Hasan, H., and Mohamad, D. (2015) Review on zinc oxide nanoparticles: antibacterial activity and toxicity mechanism. *Nano-Micro Letters*, **7**(3), 219-242.

48. Martín-Betancor, K., Aguado, S., Rodea-Palomares, I., Tamayo-Belda, M., Leganés, F., Rosal, R., and Fernández-Piñas, F. (2017) Co, Zn and Ag-MOFs evaluation as biocidal materials towards photosynthetic organisms. *Science of the Total Environment*, **595**, 547-555.

49. Tamames-Tabar, C., Imbuluzqueta, E., Guillou, N., Serre, C., Miller, S. R., Elkaïm, E., Horcajada, P., and Blanco-Prieto, M. J. (2015) A Zn azelate MOF: combining antibacterial effect. *CrystEngComm*, **17**(2), 456-462.

50. Zirehpour, A., Rahimpour, A., Arabi, A., Gh, M. S., and Soroush, M. (2017) Mitigation of thin film composite membrane biofouling via immobilizing nano-sized biocidal reservoirs in the membrane active layer mitigation of thin film composite membrane biofouling via immobilizing nano-sized biocidal reservoirs in the membrane active. *Environmental Science and Technology*, **51**(10), 5511-5522.

51. Beg, S., Rahman, M., Jain, A., Saini, S., Midoux, P., Pichon, C., Ahmad, F. J., and Akhter, S. (2017) Nanoporous metal organic frameworks as hybrid polymer-metal composites for drug delivery and biomedical applications. *Drug Discovery Today*, **22**(4), 625-637.

52. Imaz, I., Rubio-Martínez, M., An, J., Solé-Font, I., Rosi, N. L., and Maspoch, D. (2011) Metal-biomolecule frameworks (MBioFs). *Chemical Communications*, **47**(26), 7287-7302.

53. Jaros, S. W., Smoleński, P., Guedes da Silva, M. F. C., Florek, M., Król, J., Staroniewicz, Z., Pombeiro, A. J. L., and Kirillov, A. M. (2013) New silver BioMOFs driven by 1,3,5-triaza-7-phosphaadamantane-7-sulfide (PTA=S): synthesis, topological analysis and antimicrobial activity. *CrystEngComm*, **15**(40), 8060-8064.

54. Restrepo, J., Serroukh, Z., Santiago, J., Aguado, S., Gomez-Sal, P., Mosquera, M. E. G., and Rosal, R. (2017) Antibacterial Zn-MOF with hydrazinebenzoate linkers. *European Journal of Inorganic Chemistry*, **2017**(3), 574-580.

55. Duan, L.-N., Dang, Q.-Q., Han, C.-Y., and Zhang, X.-M. (2015) An interpenetrated bioactive nonlinear optical MOF containing a coordinated quinolone-like drug and Zn(ii) for pH-responsive release.

Dalton Transactions, **44**(4), 1800-1804.

56. Miller, S. R., Alvarez, E., Fradcourt, L., Devic, T., Wuttke, S., Wheatley, P. S., Steunou, N., Bonhomme, C., Gervais, C., Laurencin, D., Morris, R. E., Vimont, A., Daturi, M., Horcajada P., and Serre, C. (2013) A rare example of a porous Ca-MOF for the controlled release of biologically active NO. *Chemical Communications*, **49**(71), 7773-7775.

57. Badawi, A. M., Azzam, E. M. S., and Morsy, S. M. I. (2006) Surface and biocidal activity of some synthesized metallo azobenzene isothiouronium salts. *Bioorganic and Medicinal Chemistry*, **14**(24), 8661-8665.

58. Chen, Z.-F., Zhang, Z.-L., Tan, Y.-H., Tang, Y.-Z., Fun, H.-K., Zhou, Z.-Y., Abrahams, B. F., and Liang, H. (2008) Coordination polymers constructed by linking metal ions with azodibenzoate anions. *CrystEngComm*, **10**(2), 217-231.

59. Pettinari, C., Marchetti, F., Mosca, N., Tosi, G., and Drozdov, A. (2017) Application of metal-organic frameworks. *Polymer International*, **66**(6), 731-744.

60. Juan-Alcaniz, J., Gascon, J., and Kapteijn, F. (2012) Metal-organic frameworks as scaffolds for the encapsulation of active species: state of the art and future perspectives. *Journal of Materials Chemistry*, **22**(20), 10102-10118.

61. Meilikhov, M., Yusenko, K., Esken, D., Turner, S., Van Tendeloo, G., and Fischer, R. A. (2010) Metals@MOFs - Loading MOFs with metal nanoparticles for hybrid functions. *European Journal of Inorganic Chemistry*, (24), 3701-3714.

62. Kobayashi, H., Mitsuka, Y., and Kitagawa, H. (2016) Metal nanoparticles covered with a metal-organic framework: from one-pot synthetic methods to synergistic energy storage and conversion functions. *Inorganic Chemistry*, **55**(15), 7301-7310.

63. Yang, Q., Xu, Q., and Jiang, H.-L. (2017) Metal-organic frameworks meet metal nanoparticles: synergistic effect for enhanced catalysis. *Chemical Society Reviews*, **46**(15), 4774-4808.

64. Marambio-Jones, C., and Hoek, E. M. V. (2010) A review of the antibacterial effects of silver nanomaterials and potential implications for human health and the environment. *Journal of Nanoparticle Research*, **12**(5), 1531-1551.

65. Ximing, G., Bin, G., Yuanlin, W., and Shuanghong, G. (2017) Preparation of spherical metal - organic frameworks encapsulating ag nanoparticles and study on its antibacterial activity. *Materials Science and Engineering C*, **80**, 698-707.

66. Thakare, S. R., and Ramteke, S. M. (2017) Fast and regenerative photocatalyst material for the disinfection of E. coli from water: Silver nano particle anchor on MOF-5. *Catalysis Communications*, **102**, 21-25.

67. Mortada, B., Matar, T. A., Sakaya, A., Atallah, H., Ali, Z. K., and Karam, P. (2017) Postmetalated zirconium metal organic frameworks as a highly potent bactericide. *Inorganic Chemistry*, **56**(8), 4739-4744.

68. Guo, Y.-F., Fang, W.-J., Fu, J.-R., Wu, Y., Zheng, J., Gao, G.-Q., Chen, C., Yan, R.-W., Huang, S.-G., and Wang, C.-C. (2018) Facile synthesis of Ag@ZIF-8 core-shell heterostructure nanowires for improved antibacterial activities. *Applied Surface Science*, **435**, 149-155.

69. Tamames-Tabar, C., Cunha, D., Imbuluzqueta, E., Ragon, F., Serre, C., Blanco-Prieto, M. J., and Horcajada, P. (2014) Cytotoxicity of nanoscaled metal-organic frameworks. *Journal of Materials Chemistry B*, **2**(3), 262-271.

70. Miller, M. R., and Megson, I. L. (2009) Recent developments in nitric oxide donor drugs. *British Journal of Pharmacology*, **151**(3), 305-321.

71. Hinks, N. J., McKinlay, A. C., Xiao, B., Wheatley, P. S., and Morris, R. E. (2010) Metal organic frameworks as NO delivery materials for biological applications. *Microporous and Mesoporous Materials*, **129**(3), 330-334.

72. McKinlay, A. C., Allan, P. K., Renouf, C. L., Duncan, M. J., Wheatley, P. S., Warrender, S. J., Dawson, D., Ashbrook, S. E., Gil, B., Marszalek, B., Düren, T., Williams, J. J., Charrier, C., Mercer, D. K., Teat, S. J., and Morris, R. E. (2014) Multirate delivery of multiple therapeutic agents from metal-organic frameworks. *APL Materials*, **2**, 124108.

73. He, C., Zhang, J., and Shreeve, J. M. (2013) Dense iodine-rich compounds with low detonation pressures as biocidal agents. *Chemistry - A European Journal*, **19**(23), 7503-7509.

74. Au-Duong, A. N., and Lee, C. K. (2017) Iodine-loaded metal organic framework as growth-triggered antimicrobial agent. *Materials Science and Engineering C*, **76**, 477-482.

75. Nabipour, H., Sadr, M. H., and Bardajee, G. R. (2017) Synthesis and characterization of nanoscale zeolitic imidazolate frameworks with ciprofloxacin and their applications as antimicrobial agents. *New Journal of Chemistry*, **41**(15), 7364-7370.

76. Chowdhuri, A. R., Das, B., Kumar, A., Tripathy, S., Roy, S., and Sahu, S. K. (2017) One-pot synthesis of multifunctional nanoscale metal-organic frameworks as an effective antibacterial agent against multidrug-resistant *Staphylococcus aureus*. *Nanotechnology*, **28**(9), 095102.

77. Nabipour, H., Hossaini Sadr, M., and Rezanejade Bardajee, G. (2017) Release behavior, kinetic and antimicrobial study of nalidixic acid from [Zn$_2$(bdc)$_2$(dabco)] metal-organic frameworks. *Journal of Coordination Chemistry*, **70**(16), 2771-2784.

78. Nabipour, H., Soltani, B., and Ahmadi Nasab, N. (2018) Gentamicin loaded Zn$_2$(bdc)$_2$(dabco) frameworks as efficient materials for

drug delivery and antibacterial activity. *Journal of Inorganic and Organometallic Polymers and Materials*, **28**(3), 1206-1213.

79. Claes, B., Boudewijns, T., Muchez, L., Hooyberghs, G., Van der Eycken, E. V., Vanderleyden, J., Steenackers, H. P., and De Vos, D. E. (2017) Smart metal-organic framework coatings: triggered antibiofilm compound release. *ACS Applied Materials & Interfaces*, **9**(5), 4440-4449.

80. Lin, S., Liu, X., Tan, L., Cui, Z., Yang, X., Yeung, K. W. K., Pan, H., and Wu, S. (2017) Porous iron-carboxylate metal-organic framework: a novel bioplatform with sustained antibacterial efficacy and nontoxicity. *ACS Applied Materials & Interfaces*, **9**(22), 19248-19257.

81. Naseri, H., Sharifi, A., Ghaedi, M., Dashtian, K., Khoramrooz, S. S., Manzouri, L., Khosravanig, A. S., Pezeshkpoura, V., Sadri, F., and Askarinia, M. (2018) Sonochemical incorporated of cytosine in Cu-H$_2$bpdc as an antibacterial agent against standard and clinical strains of Proteus mirabilis with rsbA gene. *Ultrasonics Sonochemistry*, **44**, 223-230.

82. Bazer-Bachi, D., Assié, L., Lecocq, V., Harbuzaru, B., and Falk, V. (2014) Towards industrial use of metal-organic framework: Impact of shaping on the MOF properties. *Powder Technology*, **255**, 52-59.

83. Rubio-Martinez, M., Avci-Camur, C., Thornton, A. W., Imaz, I., Maspoch, D., and Hill, M. R. (2017) New synthetic routes towards MOF production at scale. *Chemical Society Reviews*, **46**(11), 3453-3480.

84. Hong, W. Y., Perera, S. P., and Burrows, A. D. (2015) Manufacturing of metal-organic framework monoliths and their application in CO$_2$ adsorption. *Microporous and Mesoporous Materials*, **214**, 149-155.

85. Valizadeh, B., Nguyen, T. N., and Stylianou, K. C. (2018) Shape engineering of metal-organic frameworks. *Polyhedron*, **145**, 1-15.

86. Rodríguez, H. S., Hinestroza, J. P., Ochoa-Puentes, C., Sierra, C. A., and Soto, C. Y. (2014) Antibacterial activity against Escherichia coli of Cu-BTC (MOF-199) metal-organic framework immobilized onto cellulosic fibers. *Journal of Applied Polymer Science*, **131**(19), 40815.

87. Wang, C., Qian, X., and An, X. (2015) In situ green preparation and antibacterial activity of copper-based metal-organic frameworks/cellulose fibers (HKUST-1/CF) composite. *Cellulose*, **22**(6), 3789-3797.

88. Neufeld, B. H., Neufeld, M. J., Lutzke, A., Schweickart, S. M., and Reynolds, M. M. (2017) Metal-organic framework material inhibits biofilm formation of *pseudomonas aeruginosa*. *Advanced Functional Materials*, **27**(34), 1702255.

89. Qian, L., Lei, D., Duan, X., Zhang, S., Song, W., Hou, C., and Tang, R. (2018) Design and preparation of metal-organic framework pa-

pers with enhanced mechanical properties and good antibacterial capacity. *Carbohydrate Polymers,* **192,** 44-51.

90. Miao, W., Wang, J., Liu, J., and Zhang, Y. (2018) Self-cleaning and antibacterial zeolitic imidazolate framework coatings. *Advanced Materials Interfaces,* **5**(14), 1800167.

91. Li, W., Zhang, Y., Li, Q., and Zhang, G. (2015) Metal-organic framework composite membranes: Synthesis and separation applications. *Chemical Engineering Science,* **135,** 232-257.

92. Mann, E. E., Rice, K. C., Boles, B. R., Endres, J. L., Ranjit, D., Chandramohan, L., Tsang, L. H., Smeltzer, M. S., Horswill, A. R., and Bayles, K. W. (2009) Modulation of eDNA release and degradation affects Staphylococcus aureus biofilm maturation. *PLoS ONE,* **4**(6), e5822.

93. Wang, J., Wang, Y., Zhang, Y., Uliana, A., Zhu, J., Liu, J., and Van der Bruggen, B. (2016) Zeolitic imidazolate framework/graphene oxide hybrid nanosheets functionalized thin film nanocomposite membrane for enhanced antimicrobial performance. *ACS Applied Materials & Interfaces,* **8**(38), 25508-25519.

94. Liu, M., Wang, L., Zheng, X., and Xie, Z. (2017) Zirconium-based nanoscale metal-organic framework/poly(ε-caprolactone) mixed-matrix membranes as effective antimicrobials. *ACS Applied Materials & Interfaces,* **9**(47), 41512-41520.

2

Effect of MOF-Graphene based Hybrid Fillers on the Properties of High Density Polyethylene Nanocomposites

Swati Singh, Muthukumaraswamy R. Vengatesan, Georgios Karanikolos and Vikas Mittal*,**

Department of Chemical Engineering, The Petroleum Institute (part of Khalifa University of Science and Technology), Abu Dhabi, UAE

Corresponding author: vik.mittal@gmail.com
**Current address*: Bletchington, Wellington County, Australia*

2.1 Introduction

Polymer nanocomposites have generated a substantial research interest owing to their promising properties as compared to conventional micro-scale nanocomposites. Polymer nanocomposites are advanced materials in which at least one of the components has a dimension in the nanoscale range (less than 100 nm). Nanofillers influence the properties of the polymers even at very low volume fractions due to their high surface area and aspect ratio [1]. Nanofillers can be divided into three different classes depending upon their dimensions. The filler materials with all dimensions in the nanometer range are known as 0-D nanofillers (e.g. silica nanoparticles) [2,3]. When two dimensions are in the nanometer range and the third is in the micrometer range, the materials are termed as 1-D nanofillers (e.g. carbon nanotubes (CNTs)) [4,5]. In case, one of the dimensions is in the nanometer range, whereas the other two are in the micrometer scale, the materials are called 2-D nanofillers (e.g. layered silicates) [6]. Overall, a large number of studies on the polymer nanocomposites with a variety of fillers have reported significant improvement in mechanical, thermal and rheological properties [7,8].

Recently, hybrid nanofillers have received extensive research attention in the field of polymer nanocomposites due to the synergetic combination of the properties of two or more different fillers. The hybrid nanofillers are made up of two or more components prepared

Metal Organic Frameworks, edited by Vikas Mittal
© 2019 Central West Publishing, Australia

using different physical and/or chemical methods. The hybridization of nanomaterials possessing different dimensions (e.g. 0-D, 1-D and 2-D) results in high performance hybrid materials as compared to the individual nanofillers, thus, leading to superior properties in the polymer nanocomposites.

Graphene contains sp^2 hybridized carbon atoms in a honeycomb structure and has been used as a nanofiller in the polymer matrices due to its superior thermal, mechanical and electrical properties [9-13]. Graphene based 2-D nanomaterials have also been commonly used with other materials for the preparation of hybrid nanofillers. Recently, Chen *et al.* [14] developed a grafted hybrid filler from CNTs and graphene platelets and incorporated it in the polyurethane matrix. It was reported that the hybrid filler displayed better dispersion than the individual fillers, which resulted in an improved electric-induced strain in the nanocomposites. Sharmila *et al.* [15] prepared epoxy nanocomposites with hybrid filler generated using reduced graphene oxide and iron nanoparticles. The improved tensile strength, dielectric properties and thermal stability were attributed to homogenous dispersion and effective interfacial adhesion owing to the hybrid filler present in the epoxy matrix.

Metal organic frameworks (MOFs) possess many superior properties, such as uniform porosity, large surface area [16] and pore volume, high thermal and chemical stability, non-toxic nature [17,18] as well as easily tunable structure, which make them efficient candidates for different applications in gas storage and separation, drug delivery [19], chemical sensing [20], catalysis [21,22], among others. MOFs, also known as coordination polymers, are composed of inorganic metal ions or metal clusters and organic linkers held together via coordination bond. Moreover, owing to the array of metal centers and functional ligands, MOF materials provide the opportunity of developing new types of composite materials with enhanced performance [23-32]. In this respect, it is of interest to study the preparation of hybrid nanofillers of MOFs with 2-D (e.g. graphene) nanomaterials via *in-situ* and *ex-situ* methods.

In the present work, the effect of the hybrid nanofillers (based on MOF and graphene) on the thermal, mechanical and rheological properties of the high density polyethylene (HDPE) matrix has been analyzed. Two different types of nanocomposites were developed: (i) composites generated with MOF coated graphene which was prepared via an *in-situ* method and (ii) composites generated using MOF and graphene via *ex-situ* method, i.e. MOF and graphene fillers were separately incorporated in the polyethylene matrix. In order to achieve homogenous dispersion of nanofillers, polyethylene-*alt*-maleic anhydride (PE-*alt*-MA) was used as a compatibilizer.

2.2 Experimental

2.2.1 Materials

HDPE, PE-*alt*-MA, xylene, dimethyl formamide (DMF), methanol, acetone, copper sulfate, trimesic acid and deionized water were procured from Sigma Aldrich. Graphite oxide (GO) was synthesized by using improved Tour's method [33].

2.2.2 Preparation of HKUST-MOF, Thermally Reduced Graphene (TRG) and HKUST-MOF/Graphene Hybrid

HKUST-MOF was prepared using the method reported in literature [34].

For the preparation of TRG, the dried GO was thermally reduced at 900° C for 30 s, using a tube furnace with nitrogen flow gas medium. The black powdered material was collected from the tube after cooling down to room temperature.

The hybrid of HKUST-MOF and graphene was prepared by one step pot method. 0.5 g of GO was mixed with 450 mL of DMF and sonicated for 1 h (solution A). Separately, 0.6 g of copper sulfate was mixed with 0.3 g of trimesic acid in 15 mL of DMF. The solution was subsequently sonicated for 5 min. 15 mL of de-ionized water (DI) was subsequently added to the solution, followed by ultra-sonication (solution B). Both solutions were subsequently mixed and kept under reflux condensation with constant stirring for 24 h. Later, the solution was filtered and washed with methanol and acetone.

2.2.3 Preparation of HDPE Nanocomposites using *Ex-situ* Fillers

0.1 g of MOF and 0.1 g of TRG were mixed using 400 mL of xylene in a round bottom flask and kept for 1 h under ultrasonication. 1 g of PE-*alt*-MA and 8.8 g of PE were then added into the solution and kept under reflux condensation for 2 h at 130° C. The solution was subsequently dried under vacuum at 100 °C to remove the solvent.

The nanocomposite test specimens were molded using mini-injection molding machine (HAAKE MiniJet PRO, Thermo Scientific) with a cylinder temperature of 180 °C, mold temperature of 125 °C, injection pressure of 430 bar for 10 s and post pressure of 500 bar for 6 s. The resultant nanocomposite was termed as PE-*ex*-MG1 (1 wt%MOF+1wt%graphene). Similarly, other composites were prepared by varying the weight percentage of the fillers as 2.5wt%MOF+2.5wt%graphene and 5wt%MOF+5wt%graphene, termed as PE-*ex*-MG2 and PE-*ex*-MG3 respectively.

2.2.4 Preparation of Nanocomposites using *In-situ* Prepared Hybrid Filler

0.2 g of the *in-situ* synthesized hybrid nanofiller was dispersed in 400 mL of xylene using ultrasonication for 1 h. Subsequently, 1 g of PE-*alt*-MA and 8.8 g of HDPE were added. The mixture was kept under reflux condensation for 2 h at 130° C. The solution was then dried under vacuum atmosphere at 100° C.

The test specimens were prepared by using the same procedure as composites with *ex-situ* fillers. The sample with 2 wt% MOF/graphene hybrid was termed as PE-*in*-MG1. Similarly, other composites were prepared by incorporating 5 wt% and 10 wt% of *in-situ* synthesized hybrid nanofiller and were termed as PE-*in*-MG2 and PE-*in*-MG3 respectively.

2.2.5 Characterization

Calorimetric properties of the materials were studied using dynamic scanning calorimetry (Discovery DSC, TA Instruments) in nitrogen atmosphere. A sample size of 3-7 mg was used for the DSC analysis. The samples were heated from 35 to 200 °C followed by cooling from 200 to -60 °C using heating and cooling rates of 10 °C/min. The second heating and cooling cycles, performed in the same temperature range, were used for the analysis of the calorimetric properties. The thermal properties of the materials were evaluated using thermogravimetric analyzer (Discovery TGA, TA instruments) with nitrogen as carrier gas. The samples were heated from 35 to 700 °C with a heating rate of 10 °C/min.

Tensile testing of the pure polymer, polymer-compatibilizer blend and composites was performed on universal testing machine (Testometric, UK). The sample dimensions for tensile test were: sample length 73 mm, gauge length 30 mm, width 4 mm and thickness 2 mm. A loading rate of 4 mm/min was used, and the tests were carried out at room temperature. Tensile modulus and yield stress were calculated using built in software Win Test Analysis. An average of five values was recorded.

The melt rheological properties (storage modulus (G'), loss modulus (G"), tan δ and complex viscosity (η^*) as a function of angular frequency (ω)) of the materials were studied using AR 2000 shear rheometer from TA Instruments (ASTM D4440). Injection molded disc shaped samples of diameter 25 mm and thickness 1.5 mm were used, and the analysis was carried out at 180 °C in parallel plate geometry of diameter 25 mm with a geometric gap of 1.4 mm. Dynamic frequency scans were performed from 0.1 to 100 rad/s at 2% strain.

X-ray diffraction (XRD) patterns of the MOF, TRG, MOF/graphene hybrid and HDPE nanocomposites were measured on a Panalytical powder diffractometer in reflection mode with CuKα radiation (λ = 1.5406 Å). The samples were scanned at room temperature from 2Θ = 5-60° with a step size of 0.02° and 10 s as step time. Zero-background holder was used to minimize the noise.

The microstructure of the samples was analyzed using transmission electron microscope (TEM) Philips CM 20 (Philips/FEI, Eindhoven) at 120 and 200 kV accelerating voltages. Thin sections of 70-90 nm thickness were microtomed from the block of the nanocomposite specimens using diamond knife at -60 °C. The thin sections were subsequently supported on 100-mesh grids sputter coated with a 3 nm thick carbon layer.

2.3 Results and Discussion

Calorimetric properties of the pure HDPE, HDPE-compatibilizer blend (10 wt%) and nanocomposites are plotted in Figure 2.1. The degree of crystallinity (X_c), peak melting temperature (T_m), peak crystallization temperature (T_c) and melting (ΔH_m)/crystallization (ΔHc) enthalpy are also listed in Table 2.1. The degree of crystallinity was determined by taking the melt enthalpy of pure crystalline HDPE as 293 J/g [28]. For pure HDPE, T_m and T_c were observed to be 135 °C and 117 °C respectively, whereas the degree of crystallinity was calculated as 56%. The melting and crystallization temperatures of HDPE did not exhibit significant change in the blend and nanocomposites. On the other hand, X_c decreased to 52% on the addition of compatibilizer (10 wt%), which might be due to the plasticization effect of the compatibilizer. Also, the incorporation of both *in-situ* prepared MOF/graphene hybrid and *ex-situ* added nanofillers to the

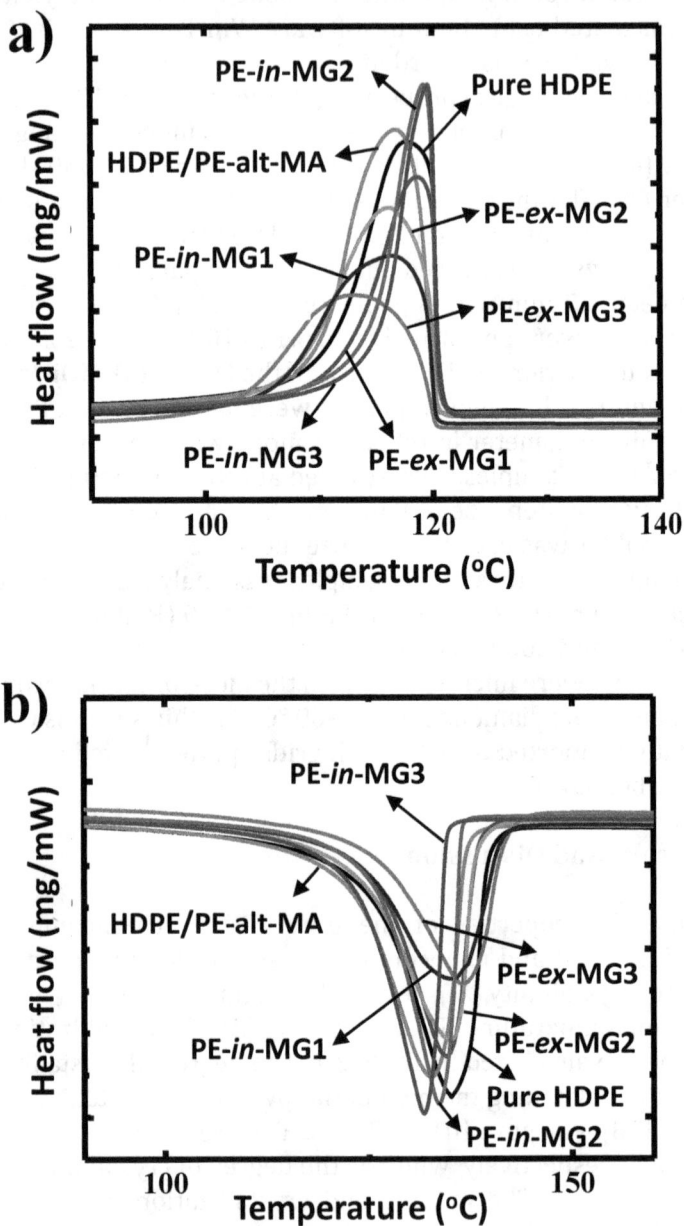

Figure 2.1 DSC (a) melting and (b) crystallization thermograms of HDPE, HDPE-compatibilizer blend and HDPE/hybrid filler nanocomposites.

Table 2.1 Calorimetric properties of HDPE, HDPE-compatibilizer blend and HDPE/hybrid filler nanocomposites

Sample	T_m(°C)	T_c(°C)	ΔH_m	ΔH_c	X_c(%)
HDPE	135	117	39.3	43.2	56
HDPE/PE-*alt*-MA	133	116	37.5	39.7	53
PE-*ex*-MG1	134	117	31.4	34.8	53
PE-*ex*-MG2	135	115	31.3	33.4	51
PE-*ex*-MG3	137	112	26.9	30.0	48
PE-*in*-MG1	136	115	28.9	32.8	47
PE-*in*-MG2	132	118	32.9	32.9	56
PE-*in*-MG3	132	119	33.5	34.0	59

HDPE matrix impacted the X_c values. In the case of *ex-situ* added nanofiller composites, X_c was observed to gradually decrease with increasing filler content. The observed decrease in the crystallinity might have resulted from the non-uniform distribution of the nanofillers. Also, as the fillers were only physically mixed, one filler might have impacted the crystallinity of the polymer more than the other. Similarly, one of the fillers might have poorer distribution than the other, thus, leading to the overall decrease in crystallinity. In the case of *in-situ* generated MOF/graphene hybrid filled nanocomposites, X_c was observed to decrease to 46% for the composite with 2 wt% filler content. However, the crystallinity increased with enhancing the filler content in the composites. For instance, for the composite with 10 wt% filler content, the crystallinity value was observed to be 59%. The initial decrease in the X_c might be due to the interference by the hybrid nanofiller and compatibilizer with the arrangement of chain molecules in the crystal lattice of the polymer. Further addition of the hybrid filler in the HDPE matrix exhibited a gradual nucleating effect which resulted in improved X_c values.

The thermal properties of the materials were examined under nitrogen atmosphere to ascertain the onset degradation temperature and char yield. Table 2.2 lists T^d_{10} (°C) and T^d_{40} (°C) values (temperatures at which 10 and 40% weight loss has taken place, respectively) as well as char yield at 700 °C (% of non-volatile material present at 700 °C). Figure 2.2 also presents the TGA thermograms of the materials. As observed from the table, the initiation of degradation (10% weight loss) temperature was observed to be higher for pure HDPE as compared to the blend and nanocomposites, probably due to the presence of the compatibilizer. On the other hand, the temperatures for 40 wt% degradation were observed to be slightly higher in the nanocomposites. It indicated that the nanofillers did not significantly impact the thermal behavior of the polymer, and the polymer still retained high temperature stability. The char yield was observed to increase with the amount of filler due to the increment in the fraction of the non-volatile material.

Table 2.2 Thermal stability and char yield of HDPE, HDPE-compatibilizer blend and HDPE/hybrid filler nanocomposites

Sample	$T^d_{10}(^{\circ}C)$	$T^d_{40}(^{\circ}C)$	Char yield at $700^{\circ}C$ (%)
Pure HDPE	445	472	0.00
HDPE/PE-alt-MA	443	477	0.25
PE-*ex*-MG1	446	479	1.77
PE-*ex*-MG2	430	479	3.96
PE-*ex*-MG3	431	477	7.97
PE-*in*-MG1	426	477	1.89
PE-*in*-MG2	427	479	2.83
PE-*in*-MG3	423	479	5.54

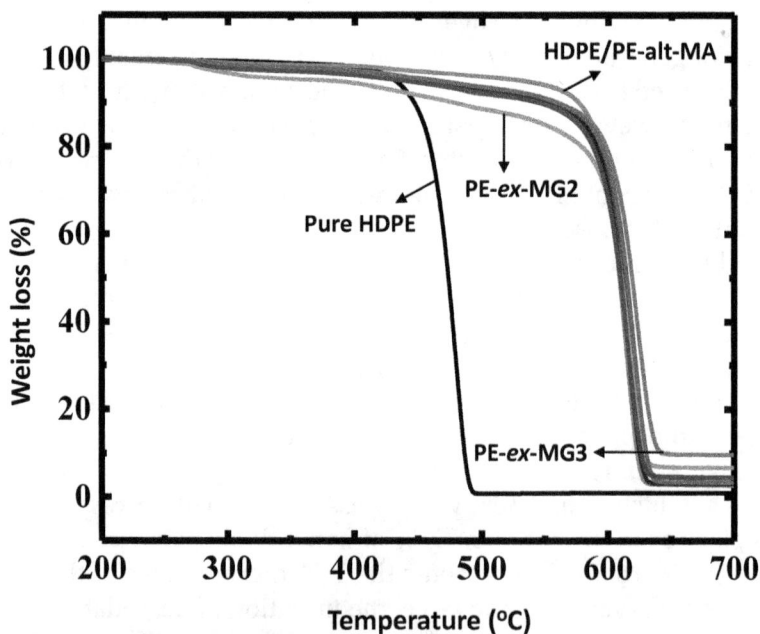

Figure 2.2 TGA curves of HDPE, HDPE-compatibilizer blend and HDPE/ hybrid filler nanocomposites.

The effect of the *in-situ* hybrid nanofiller and *ex-situ* mixed fillers on the mechanical properties of the materials has been depicted in Table 2.3. The addition of 10 wt% compatibilizer to the HDPE matrix resulted in a reduction in the tensile modulus due to the plasticization effect. The *in-situ* hybrid MOF/graphene incorporated nanocomposites showed a gradual increase in the modulus with enhancement in the hybrid filler content. In case of *ex-situ* mixed filler system, the addition of nanofillers also increased the modulus for the composite with 2 wt% filler content. However, the modulus was observed to gradually decrease with increase in the filler content. It indicated that the *in-situ* hybrid fillers imparted superior mechanical performance to the polymer. In the case of *in-situ* hybrid filler, MOF molecules are expected to be uniformly stacked on the surface of the graphene sheets which reduces sheet stacking, whereas in the case of *ex-situ* method, the individual filler materials may have distributed in different regions in the polymer matrix, thus, resulting in less optimal enhancement in the mechanical performance (Figure 2.3). The superior performance of *in-situ* hybrid filler system can also be gauged from the simultaneous increment in the tensile strength of the polymer, which was observed to decrease in the case of *ex-situ* added fillers. For instance, the *in-situ* hybrid filler based polymer nanocomposite with 10 wt% filler fraction exhibited higher tensile strength (36 MPa) than the pure polymer (34 MPa), whereas the same nanocomposite with *ex-situ* filler addition exhibited a much lower value of 24 MPa. The strain at break was observed to decrease in both *in-situ* and *ex-situ* systems, however, the reduction was more severe in the case of *ex-situ* added filler, thus, exhibiting higher extent of brittleness.

Table 2.3 Mechanical properties of HDPE, HDPE-compatibilizer blend and HDPE/hybrid filler nanocomposites

Sample	UTS (MPa)	Modulus (MPa)	Strain at Break (%)
Pure HDPE	34	598	31
HDPE/PE-alt-MA	31	567	41
PE-*ex*-MG1	32	730	11
PE-*ex*-MG2	27	707	10
PE-*ex*-MG3	24	634	4
PE-*in*-MG1	32	710	17
PE-*in*-MG2	32	756	15
PE-*in*-MG3	36	798	9

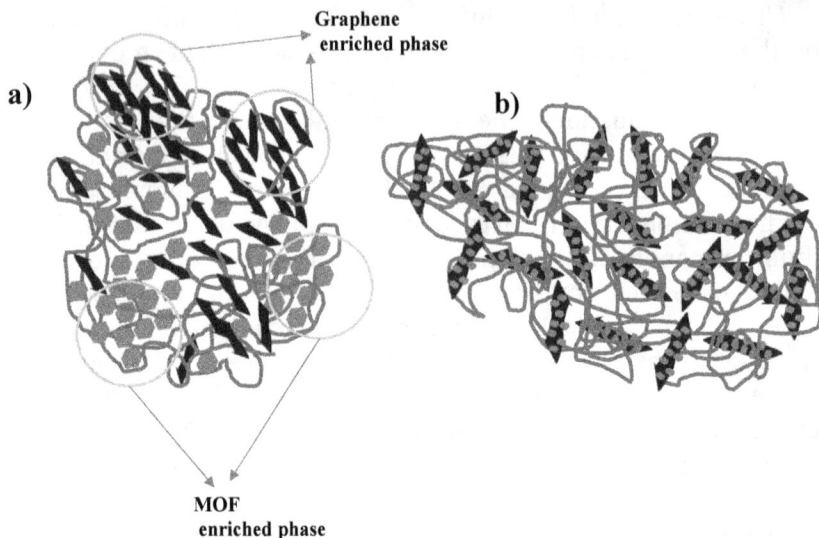

Figure 2.3 Schematic representation of a) *ex-situ* hybrid filler nanocomposites and b) *in-situ* hybrid filler nanocomposites.

The effect of hybrid fillers on the polymer chain motion, filler dispersion and polymer-filler interactions has been studied by the melt rheological analysis [35,36]. The shape of nanofiller has been reported to influence the melt rheological properties of the nanocomposites [37]. Figure 2.4a,b shows the plots of storage (G') and loss modulus (G") as a function of angular frequency. The G' and G" values of the HDPE matrix and nanocomposites increased with frequency. The nanofillers had an insignificant impact on the storage and loss modulus of the polymer. Also, for both types of filler systems, a well-defined plateau of G' was observed at lower frequency in the composites, indicating a transition from liquid-like to pseudo solid-like viscoelastic behavior leading to a percolation network structure [38,39]. The miscibility analysis using the rheological data also indicated higher extent of compatibility in the case of *in-situ* filler system, thus, confirming the higher extent of property enhancement as compared to *ex-situ* system. Figure 2.4c also presents the complex viscosity of the materials as a function of angular frequency. The nanocomposites showed a frequency independent Newtonian flow like behavior at lower frequencies (less than 1 rad/s), followed by a shear thinning behavior [40,41]. As the viscosity of the polymer remained largely unaffected by the addition of nanofillers, the composites could be processed using the same processing conditions as the pure polymer.

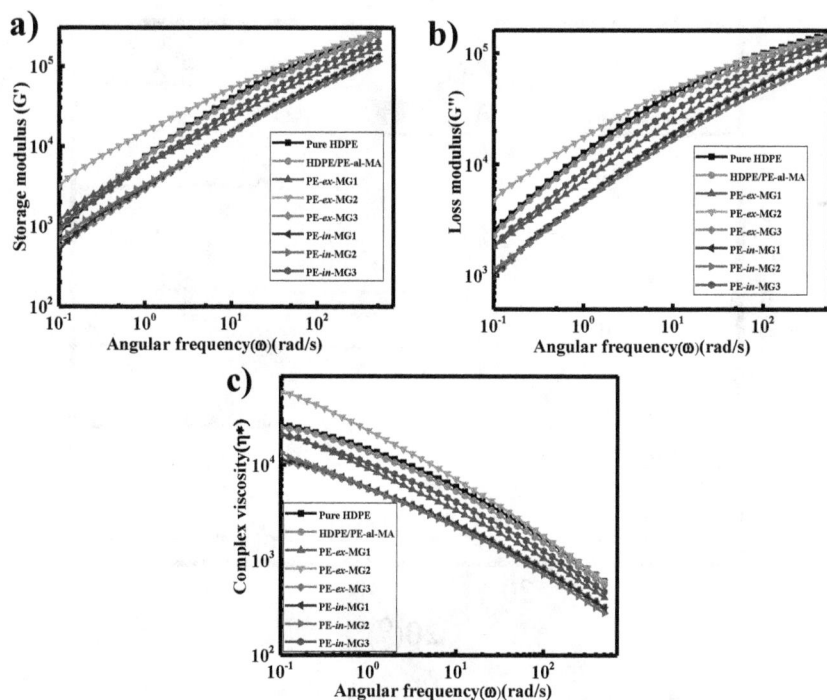

Figure 2.4 Melt rheology of HDPE, HDPE-compatibilizer blend and HDPE/hybrid filler nanocomposites: a) storage modulus, b) loss modulus and c) complex viscosity as a function of angular frequency.

XRD was performed on the pristine polyethylene and nanocomposites to study the changes in the crystalline structure with filler incorporation, as shown in Figure 2.5. High density polyethylene exhibited two diffraction peaks at $2\theta = 22.4°$ and $24.7°$ corresponding to (110) and (200) planes. It indicated the orthorhombic crystal structure of polyethylene, as reported in literature [25]. Other weak reflections were observed at $30°$ and $37.1°$ for pristine polyethylene corresponding to (210) and (020) reflection planes respectively [26,27]. No noticeable change in the characteristic diffraction peaks of (110) and (200) planes was observed in the nanocomposites, thus, indicating that the incorporation of nanofillers did not affect the crystal structure of the polymer.

The TEM micrographs of the *in-situ* generated filler and composites with 5 wt% *in-situ* and *ex-situ* filler systems are presented in Figure 2.6. The TEM micrograph of the *in-situ* filler system in Figure 2.6a showed uniform stacking of MOF on the graphene platelets, which

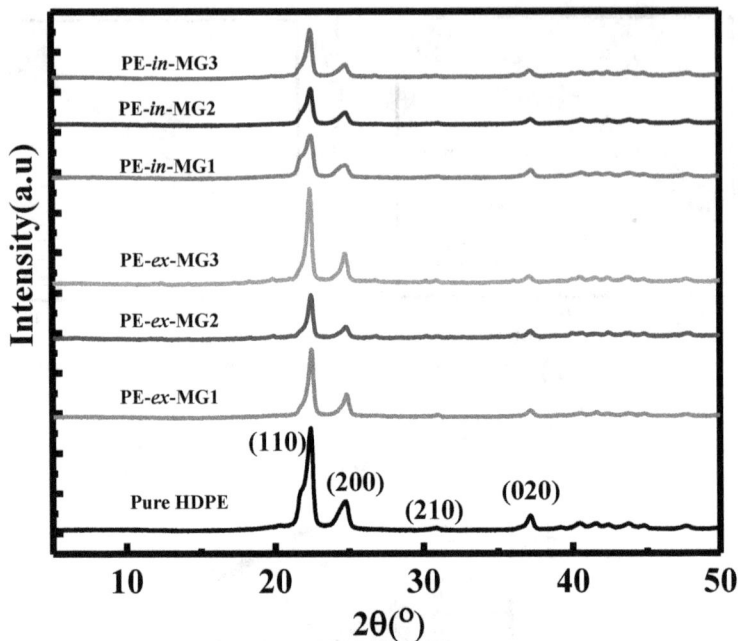

Figure 2.5 XRD of HDPE and HDPE/hybrid filler nanocomposites.

helped to enhance sheet delamination and reduce filler aggregation in the polymer matrix. The TEM image of PE-*ex*-MG2 composite in Figure 2.6b presents the enriched phase of MOF in the polymer region, along with a small degree of agglomeration. It confirmed the

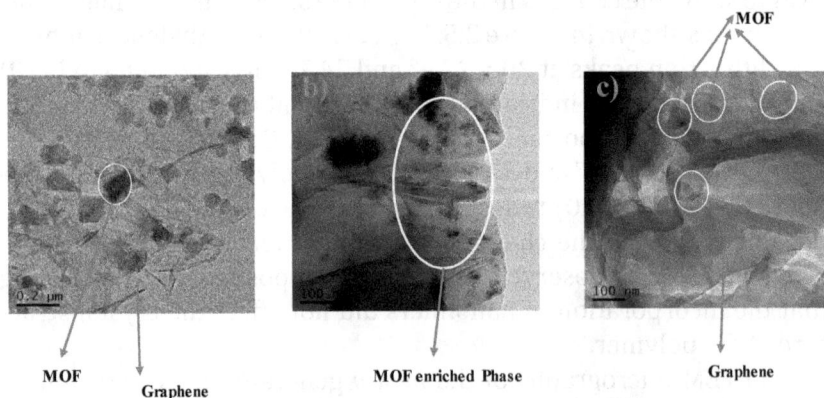

Figure 2.6 TEM images of a) *in-situ* MOF coated graphene, b) PE-*ex*-MG2 and c) PE-*in*-MG2.

earlier notion that the filler phases in the *ex-situ* system may be dispersed at different locations in the polymer matrix. In the TEM micrograph of the *in-situ* hybrid filler nanocomposite (PE-*in*-MG2) (Figure 2.6c), a uniform dispersion of nanofiller in the polymer phase was observed, which can be attributed to the relatively better property profiles as compared to the composites with *ex-situ* added fillers.

2.4 Conclusion

In this work, the effect of hybrid nanofillers composed of MOF and graphene was analyzed on the properties of the polyethylene matrix. The hybrid fillers were generated either *in-situ* prior to the incorporation in the polymer matrix or were added individually to the polymer matrix (*ex-situ*). *In-situ* hybrid nanofiller based nanocomposites exhibited higher modulus and strength than the *ex-situ* hybrid nanofiller based composites. The degree of crystallinity of the *in-situ* hybrid nanofiller based nanocomposites also showed an improvement with filler content. The thermal and rheological properties of the polymer were not significantly affected by the presence of the hybrid nanofillers. The morphological analysis confirmed the uniform filler distribution in the case of *in-situ* hybrid filler based composites, thus, indicating the advantage of hybridizing the fillers prior to the incorporation in the polymer matrix.

References

1. Brechet, Y., Cavaille, J. Y., Chabert, E., Chazeau, L., Dendievel, R., Flandin, L., and Gauthier, C. (2001) Polymer based nanocomposites: effect of filler-filler and filler-matrix interactions. *Advanced Engineering Materials*, **3**(8), 571-577.
2. Mark, J. E. (1996) Ceramic-reinforced polymers and polymer-modified ceramics. *Polymer Engineering and Science*, **36**(24), 2905-2920.
3. Chang, W. Y., Lee, K. J., and Nam, J. D. (2003) Reactive dispersion and mechanical property of dicyanate/montmorillonite nanocomposite. *Polymer Korea*, **27**, 75-83.
4. Calvert, P. (1997) Potential applications of nanotubes. In: *Carbon Nanotubes: Preparation and Properties*, Ebbesen, T. W. (ed.), CRC Press, USA, pp. 277-292.
5. Favier, V., Canova, G. R., Shrivastava, S. C., and Cavaille, J. Y. (1997) Mechanical percolation in cellulose whisker nanocomposites. *Polymer Engineering and Science*, **37**, 1732-1739.

6. Mittal, V. (2006) *Organic Modifications of Clay and Polymer Surfaces for Specialty Applications*, PhD Thesis, ETH Zurich, Switzerland.

7. Mittal, V. (2007) polypropylene-layered silicate nanocomposites: filler matrix interactions and mechanical properties. *Journal of Thermoplastic Composite Materials,* **20**, 575-599.

8. Chaudhry, A. U., and Mittal, V. (2013) High-density polyethylene nanocomposites using masterbatches of chlorinated polyethylene/graphene oxide. *Polymer Engineering and Science,* **53**, 78-88.

9. Neto, A. H. C., Guinea, F., Peres, N. M. R., Novoselov, K. S., and Geim, A. K. (2009) The electronic properties of graphene. *Reviews of Modern Physics,* **81**, 109-162.

10. Katsnelson, M. I., Novoselov, K. S., and Geim A. K. (2006) Chiral tunneling and the Klein paradox in graphene. *Nature Physics,* **2**, 620-625.

11. Zhang, Y. B., Tan, Y. W., Stormer, H. L., and Kim P. (2005) Experimental observation of the quantum Hall effect and Berry's phase in graphene. *Nature,* **438**, 201-204.

12. Reiner-Rozman, C., Larisika, M., Nowak, C., and Knoll, W. (2015) Graphene-based liquid-gated field effect transistor for biosensing: theory and experiments. *Biosensors and Bioelectronics,* **70**, 21-27.

13. Novoselov, K. S., Geim, A. K., Morozov, S. V., Jiang, D., Zhang, Y., Dubonos, S. V., Grigorieva I. V., and Firsov, A. A. (2004) Electric field effect in atomically thin carbon films. *Science,* **306**, 666-669.

14. Chen, T., Pan, L., Lin, M., Wang, B., Liu, L., Li, Y., Qiu, J., and Zhu, K. (2015) Dielectric, mechanical and electro-stimulus response properties studies of polyurethane dielectric elastomer modified by carbon nanotube-graphene nanosheet hybrid fillers. *Polymer Testing,* **40**, 4-11.

15. Sharmila, T. K. B., Antony, J. V., Jayakrishnan, M. P., Beegum, P. M. S., and Thachil, E. T. (2015) Mechanical, thermal and dielectric properties of hybrid composites of epoxy and reduced graphene oxide/iron oxide. *Materials and Design,* **90**, 66-75.

16. Chae, H. K., Siberio-Pérez, D. Y., Kim, J., Go, Y., Eddaoudi, M., Matzger, A. J., O'Keeffe, M., and Yaghi, O. M. (2004) A route to high surface area porosity and inclusion of large molecules in crystals. *Nature,* **427**, 523-527.

17. Furukawa, H., Cordova, K. E., O'Keeffe, M., and Yaghi, O. M. (2013) The chemistry and applications of metal-organic frameworks. *Science,* **341**(6149), 1230444.

18. Tanabe, K. K., and Cohen, S. M. (2011) Post synthetic modification of metal-organic frameworks-a progress report. *Chemical Society Review,* **40**, 498-519.

19. Cook, T. R., Zheng, Y. R., and Stang, P. J. (2013) Metal-organic frameworks and self-assembled supramolecular coordination complexes: Comparing and contrasting the design, synthesis, and functionality

of metal-organic materials. *Chemical Reviews,* **113**, 734-777.

20. Cheetham, A. K., Rao, C. N., and Feller, R. K. (2006) Structural diversity and chemical trends in hybrid inorganic-organic framework materials. *Chemical Communications,* **46**, 4780-4795.
21. Vaesen, S., Guillerm, V., Yang, Q., Wiersum, A. D., Marszalek, B., Gil, B., Vimont, A., Daturi, M., Devic, T., Llewellyn, P. L., Serre, C., Maurin, G., and Weireld, G. D. (2013) A robust amino-functionalized titanium(IV) based MOF for improved separation of acid gases. *Chemical Communications,* **49**, 10082-10084.
22. Hamilton, T. D., Papaefstathiou, G. S., and MacGillivray, L. R. (2005) Template-controlled reactivity: Following nature's way to design and construct metal-organic polyhedra and polygons. *Journal of Solid State Chemistry,* **178**, 2409-2413.
23. Anbia, M., and Hoseini, V. (2012) Development of MWCNT@MIL-101 hybrid composite with enhanced adsorption capacity for carbon dioxide. *Chemical Engineering Journal,* **191**, 326-330.
24. Buso, D., Jasieniak, J., Lay, M. D. H., Schiavuta, P., Scopece, P., Laird, J., Amenitsch, H., Hill, A. J., and Falcaro, P. (2012) Highly luminescent metal-organic frameworks through quantum dot doping. *Small,* **8**, 80-88.
25. Jahan, M., Bao, Q., and Loh, K. P. (2012) Electrocatalytically active graphene-porphyrin MOF composite for oxygen reduction reaction. *Journal of the American Chemical Society,* **134**, 6707-6713.
26. Xu, Y., Wen, Y., Zhu, W., Wu, Y.-n., Lin, C., and Li, G. (2012) Electrospun nanofibrous mats as skeletons to produce MOF membranes for the detection of explosives. *Materials Letters,* **87**, 20-23.
27. Prasanth, K. P., Rallapalli, P., Raj, M. C., Bajaj, H. C., and Jasra, R. V. (2011) Enhanced hydrogen sorption in single walled carbon nanotube incorporated MIL-101 composite metal-organic framework. *International Journal of Hydrogen Energy,* **36**, 7594-7601.
28. Yang, Y., Ge, L., Rudolph, V., and Zhu, Z. (2014) In situ synthesis of zeolitic imidazolate frameworks/carbon nanotube composites with enhanced CO_2 adsorption. *Dalton Transactions,* **43**, 7028-7036.
29. Lee, H., Choi, Y. N., Choi, S. B., Seo, J. H., Kim, J., Cho, I. H., Gang, S., and Jeon, C. H. (2014) In situ neutron powder diffraction and X-ray photoelectron spectroscopy analyses on the hydrogenation of MOF-5 by Pt-doped multiwalled carbon nanotubes. *Journal of Physical Chemistry C,* **118**, 5691-5699.
30. Yang, S. J., Choi, J. Y., Chae H. K., Cho, J. H., Nahm, K. S., and Park, C. R. (2009) Preparation and enhanced hydrostability and hydrogen storage capacity of CNT@MOF-5 hybrid composite. *Chemistry of Materials,* **21**, 1893-1897.
31. Lu, G., Li, S., Guo, Z., Farha, O. K., Hauser, B. G., Qi, X., Wang, Y., Wang, X., Han, S., Liu, X., DuChene, J. S., Zhang, H., Zhang, Q., Chen, X., Ma, J., Loo, S. C. J., Wei, W. D., Yang, Y., Hupp, J. T., and Huo, F. (2012) Imp-

arting functionality to a metal-organic framework material by controlled nanoparticle encapsulation. *Nature Chemistry,* **4**, 310-316.

32. Mao, Y., Li, J., Cao, W., Ying, Y., Hu, P., Liu, Y., Sun, L., Wang, H., Jin, C., and Peng, X. (2014) General incorporation of diverse components inside metal-organic framework thin films at room temperature. *Nature Communications,* **5**, 5532.

33. Marcano, D. C., Kosynkin, D. V., Berlin, J. M., Sinitskii, A., Sun, Z., Slesarev, A., Alemany, L. B., Lu, W., and Tour, J. M. (2010) Improved Synthesis of Graphene Oxide. *ACS Nano,* **4**, 4806-4814.

34. Ke, F., Qiu, L. G., Yuan, Y. P., Peng, F. M., Jiang, X., Xie, A. J., Shen Y. H., and Zhu, J. F. (2011) Thiol-functionalization of metal-organic framework by a facile coordination-based post synthetic strategy and enhanced removal of Hg^{2+} from water. *Journal of Hazardous Materials,* **196,** 36-43.

35. Bousmina, M. (2006) Study of intercalation and exfoliation processes in polymer nanocomposites. *Macromolecules,* **39**, 4259-4263.

36. Cassagnau, P. (2008) Melt rheology of organoclay and fumed silica nanocomposites. *Polymer,* **49**, 2183-2196.

37. Knauert, S. T., Douglas, J. F., and Starr, F. W. (2007) The effect of nanoparticle shape on polymer-nanocomposite rheology and tensile strength. *Journal of Polymer Science, Part B: Polymer Physics,* **45**, 1882-1897.

38. Achaby, M. E., and Qaiss, A. (2013) Processing and properties of polyethylene reinforced by graphene nanosheets and carbon nanotubes. *Materials and Design,* **44**, 81-89.

39. Horst, M. F., Quinzani, L. M., and Failla, M. D (2014) Rheological and barrier properties of nanocomposites of HDPE and exfoliated montmorillonite. *Journal of Thermoplastic Composite Materials,* **27**, 106-125.

40. Li, Y., Zhu, J., Wei, S., Ryu, J., Sun, L., and Guo, Z. (2011) Poly (propylene)/graphene nanoplatelet nanocomposites: Melt rheological behavior and thermal, electrical, and electronic properties. *Macromolecular Chemistry and Physics,* **212**, 1951-1959.

41. Essabir, H., Rodrigue, D., Bouhfid, R., and Qaiss, A. K. (2018) Effect of nylon 6 (PA6) addition on the properties of glass fiber reinforced acrylonitrile-butadiene-styrene. *Polymer Composites,* **39**(1), 14-21.

3

Application of Metal Organic Frameworks in Electrochemical Sensors for Biomolecules Detection

Hongli Zhao, Libo Shi, Xuan Cai and Minbo Lan*

Shanghai Key Laboratory of Functional Materials Chemistry, School of Chemistry and Molecular Engineering, East China University of Science and Technology, Shanghai, 200237, People's Republic of China

Corresponding author: minbolan@ecust.edu.cn

3.1 Introduction

Today, one of the most challenging problems in our society is the public health intimidated by various diseases. For example, as reported by World Health Organization (WHO) in 2012, nearly 35.6 million people live with Alzheimer's disease. This number is expected to double by 2030 (65.7 million). Another example is the data about cancer; according to WHO in 2017, nearly 1 in 6 deaths globally is due to cancer. Under these circumstances, it is of significant importance to realize the pathological research, early diagnostic and continuous monitoring of the diseases. From the perspective of chemistry, it is urgent to uncover specific information, such as functions, metabolisms as well as spatial and temporal distributions, of small biological molecules or protein biomarkers specifically related to the common diseases.

To date, various methodologies have been developed to obtain spatial and temporal information of the biomolecules in order to achieve their detection. These methodologies include chemiluminescence, fluorescence, surface enhanced Raman spectroscopy, etc., and have been widely reported by many research groups [1-3]. As an alternative to these spectroscopy methodologies, electrochemical sensors have also been widely explored for their application in biomolecules detection [4,5]. With the development of novel fabrication and device technologies, electrochemical sensors show the superiority of low-cost, easy-operation and fast-detection in most cases. Moreover,

Metal Organic Frameworks, edited by Vikas Mittal
© 2019 Central West Publishing, Australia

they are also qualified for meeting the demands of analytical performance with high sensitivity, selectivity and stability.

Although carbon-, silver-, gold- or platinum-bare electrodes can be directly applied for electrochemical analysis in some cases, however, for the electrochemical detection of biomolecules, electrode modification with nanomaterials plays a critical role to improve the detection performance [6-9]. The performance of the nanomaterials-based sensors can be improved from the following aspects: (1) nanomaterials modified electrodes are efficient electrocatalysts for some molecules, thus, providing the basis for obtaining the electrochemical signals directly from the targets via electro-oxidation/reduction, (2) in some cases, conductive nanomaterials such as carbon nanotubes, graphene and some polymers can facilitate the electron transfer from the electrochemical reaction interface to electrode, thus, amplifying the electrochemical signals and (3) for biosensors, the immobilized biological active substances like enzymes and antibodies are easy to be denatured in most cases by the environmental temperature and pH. Nanomaterials with porous structure can provide a friendly environment to protect these biological substances, thus, improving the detection stability. From these considerations, one of the research directions on electrochemical sensors is the exploration of novel nanomaterials to improve the detection performance for various analytes.

Based on the above discussion, one of the research directions for electrochemical sensors is the exploration of novel nanomaterials to improve the detection performance for various analytes. Metal organic frameworks (MOFs), also known as porous coordination polymers (PCPs), are a new kind of materials generated by the coordination between metal ions and organic ligands. The topological structure of MOFs lead to the ultra-high BET surface area [10,11] and porosity with up to 90% free volume [12]. Moreover, the large number of available metal ions and organic ligands lead to the high tunability of MOFs designs through facile synthesis. With these unique properties, MOFs have been explored for their applications in storage and separation [13,14], drug delivery [15-17] and heterogeneous catalysis [18-20]. Recently, use of MOFs in electrochemical sensors has become a topic of intense research interest. The multiple choices of the composition, large specific surface area, tunable morphology and porosity of MOFs provide an unlimited potential for the utilization in electrochemical sensors for biomolecules detection. The increasing trend can be observed from the number of published studied in recent years, as shown in Figure 3.1. There have also been a number of

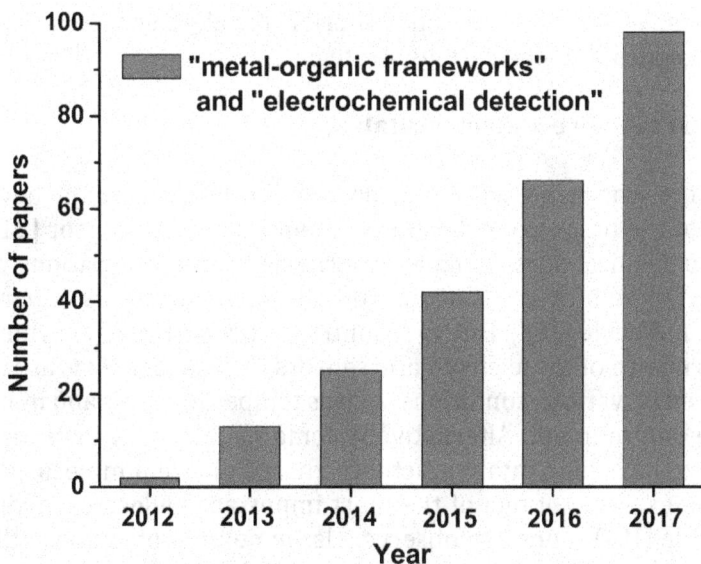

Figure 3.1 The number of research articles relevant to MOFs application in electrochemical detection. Data was collected from Web of Science on January 19, 2018 by searching the terms "metal-organic frameworks" and "electrochemical detection" as theme.

reviews, mainly on the synthesis and applications of MOFs. However, limited number of reviews are available concerning MOFs-based electrochemical sensors despite the increasing number of research studies. Herein, we focus our attention on the applications of MOFs in electrochemical sensors for the detection of biomolecules. The literature survey on MOFs-based electrochemical sensors reveals that the functions of MOFs in electrochemical sensors can be summarized into the following three aspects: (1) MOFs can be directly applied as electrocatalysts for the direct catalytic detection of some biomolecules, (2) MOFs can be utilized as calcination precursors for some porous carbon, metal oxide and their composites, which can be electrocatalysts with high conductivity and activity and (3) MOFs with unique porosity and specific surface area also present an outstanding immobilization platform for some biorecognition units and nanomaterials. In this chapter, we have tried to elaborate these functions of MOFs in the generation of electrochemical sensors for the detection of biomolecules. Any oversight of the studies related to the applications of MOFs due to the continuous development is purely unintentional. It is envisaged that this work will be a useful contribution to this research field and will play an active role in guiding the design and synthesis of MOFs-based nanomaterials for electrochemical sensing of biomolecules using various strategies.

3.2 MOFs as Direct Electrocatalysts

Since the pioneering work of Lyon and Clark leading to the development of the first enzymatic electrochemical glucose sensor [21], the electrochemical sensors have experienced three generations of development, which are based on the enzymatic catalysis toward the target molecules [22]. Due to the intrinsic properties of enzymes, the performance of these enzymatic sensors is unsatisfactory, as it is influenced by various conditions such as temperature, pH and moisture in the environment. Alternatively, some metallic materials are also found to have high catalytic activity toward different molecules. Taking the example of one of the most important molecules, hydrogen peroxide (H_2O_2), one can observe a large number of studies utilizing various metallic materials such as gold [23,24], silver [25,26], platinum [27,28], copper [29,30], iron [31,32], nickel [33,34] to accomplish the non-enzymatic electroanalysis of target H_2O_2. MOFs are composed of repeated metal complex units with metal ions coordinated by organic linkers. The most common metallic species contained in of MOFs are the transition metals such as Cu, Co, Fe, Ni, etc., which are redox active under most circumstances. Based on these facts, MOFs with redox-active metallic species can become the promising candidates as MOFzymes (the term "MOFzymes" was firstly proposed by Li *et al.* [35] when they found the intrinsic protease-like activity of Cu-MOF in 2014) with intrinsic enzyme mimicking activity. These facts provide a basis for the electroanalysis of the targets by using MOFs as direct electrocatalysts. More importantly, the tunability of the structure and properties of MOFs gives them more promising potential in direct electrocatalysis of the analytes.

One of the most pioneering works using MOFs as direct electrocatalysts was performed by Pang *et al.* [36]. In the presence of cetyltrimethylammonium bromide (CTAB) surfactant, the authors synthesized $Cu_3(btc)_2$ (btc^{3-} = benzene-1,3,5-tricarboxylate) nano/microcrystals with controlled morphology including nano-cube, truncated cube, cuboctahedron and octahedron for non-enzymatic glucose sensing. The authors found that the nano-cube structure with (1) nanopores on (100) crystal planes and (2) smaller size for shorter diffusion length exhibited the highest electro-catalytic activity toward glucose among MOFs crystals. Zhang *et al.* [37] synthesized zeolitic imidazolate framework-67 (ZIF-67, a cobalt-based MOFs) with micro-flowers morphology for both supercapacitor and electroanalysis application [38]. ZIF-67 micro-flowers showed outstanding electro-reduction activity toward H_2O_2 in alkaline media, and it was further applied for the non-enzymatic H_2O_2 sensing [37]. Yuan and co-workers proposed Co-based MOFs as the electrode materials for the electrocatalytic oxidation of reduced glutathione (GSH) [38]. A simple solvothermal synthesis was carried out to obtain [Co(tib)$_2$]·2NO$_3$ (tib = 1,3,5-tris(1-imidazolyl) benzene). Low detection limit toward GSH electrochemical detection was obtained.

Although the high tunability of the morphology, crystal plane and metal ions composition in MOFs makes them an outstanding choice of electrocatalysts for various biomolecular analytes, there are still some critical issues in their direct application which need to be resolved. For instance, the electrochemical active sites are buried inside the crystalline structure, so the target molecules cannot efficiently diffuse to the active sites in the MOFs crystals. Despite the porous feature of MOFs, the micropores (in most cases) cannot effectively improve the low response kinetics. One more critical issue is the low conductivity. Since most of the common organic ligands in MOFs are electrochemically inert, MOFs are often not conductive. This feature results in a significant obstruction to the electron transfer between the reaction interface and electrode. To resolve the conductivity limitation, one of the effective strategies is the use of the heterogeneous conductive carbon materials. For example, Bao *et al.* [39] have developed a method for incorporating multi-walled carbon nanotubes (NWCNTs) into a manganese-based MOFs, Mn-BDC (BDC = benzenedicarboxylic acid). The CNTs were observed to split the bulky Mn-BDC into thin layers, thus, boosting the conductivity and electrocatalytic ability. The resulting Mn-BDC@MWCNTs were used as simultaneous electroanalysis of ascorbic acid (AA), dopamine (DA) and uric acid (UA) in body fluids [39]. By using CNTs as heterogeneous incorporation materials, the group also anchored Ni(II)-MOF nanoparticles with a size of 2~4 nm onto MWCNTs to accomplish high sensing performance toward H_2O_2 with the synergistic catalytic activity between Ni(II)-MOF nanoparticles and MWCNTs [40]. Moreover, other conductive materials like graphene [41] have also been explored for improving the electroanalysis performance of MOF electrocatalysts. Wang *et al.* [42] incorporated graphene oxide (GO) with HKUST-1 for the detection of H_2O_2 in biological samples as shown in Figure 3.2A. GO in this work acted as structure-directing agent, which induced and participated in the formation of HKUST-1 with flower shape. The formed graphene and HKUST-1 composites showed high detection performance toward H_2O_2 in human serum and living RAW 264.7 cell fluids due to the increased surface area, electronic conductivity and redox-activity of the material [42].

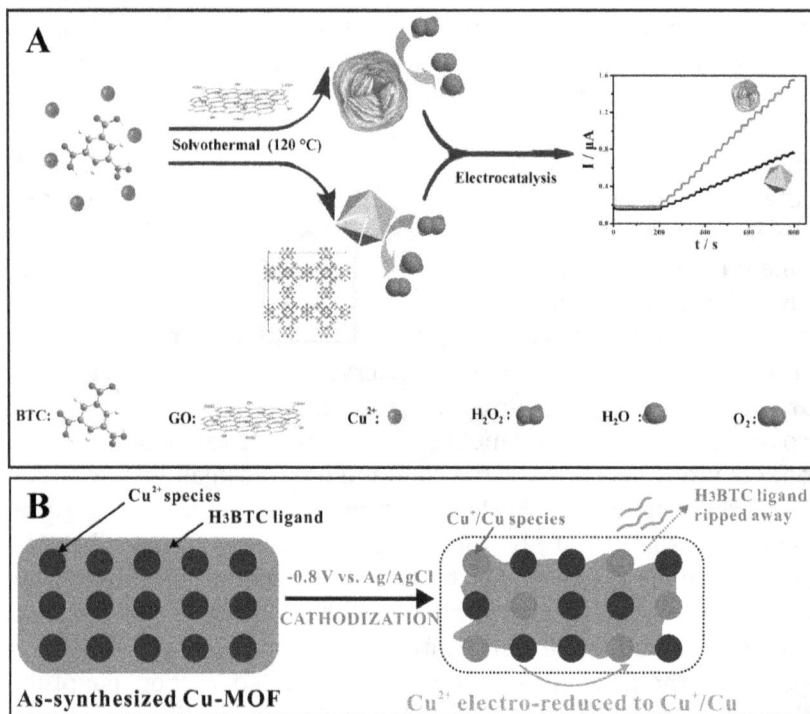

Figure 3.2 (A) Schematic illustration of the preparation of SGO@HKUST-1 and its application as a non-enzymatic electrocatalyst for H_2O_2. Reproduced from Reference [42] with permission from American Chemical Society. (B) Schematic illustration of the evolution mechanism of Cu-MOF induced by cathodization treatment for achieving the improved electrocatalytic activity. Reproduced from Reference 46 with permission from Wiley.

An alternative strategy to improve the electron transfer capability of MOFs electrocatalysts is doping redox active species into MOFs instead of inert organic ligands. This can be accomplished by either immobilizing ferrocene into MOFs [43], or applying pre-designed redox active ligands to coordinate with metal ions [44,45]. Interestingly, this mode can be regarded as a "built-in second generation electrochemical sensor", in which metals in MOFs electrocatalyze the analyte and the electrons can be mediated by redox active ligands.

The third method to resolve the aforementioned problem has been proposed by our group [46]. As shown in Figure 3.2B, by applying a cathodization treatment on Cu-MOF, Cu ions in the MOFs are reduced to Cu(0), forming Cu nanoparticles. The inert organic ligands can be removed partly during this procedure. This possible mechanism was demonstrated via a series of characterizations including crystal phases, surface compositions and morphologies. Results show that using the cathodization treatment, the non-enzymatic sensing response of the Cu-MOF based structures toward glucose could be improved by 5 times.

3.3 MOFs as Calcination Precursors

The second main application of MOFs in electroanalysis is the use as calcination precursors. With the thermal treatment of MOFs, a variety of metal and carbon-based hybrid nanomaterials can be synthesized with distinct physical and chemical properties. In general, the approach of thermal treatment of MOFs shows some advantages: (1) with the diverse composition of metal ions and organic ligands in MOFs, the calcination process with flexible conversion conditions can lead MOFs into various nanohybrids with desired metal and carbon composition. (2) MOFs can serve as self-sacrificial templates, significantly reducing the aggregation during the calcination. The robustness of the frameworks can also avoid the collapse of the structure. (3) The resulting nanohybrid generally exhibits large specific surface area and interconnected pores, which are very crucial for the improved electrochemical performance. By taking full advantage of these features, MOFs-derived nanomaterials are widely explored for their application in the energy storage fields [47-50]. Recently, research on the MOFs-derived nanomaterials toward electrochemical sensors applications has also emerged. Metal and carbon nanohybrids obtained by calcination of MOFs precursors with the feature of large surface area and porosity make significant contribution to the high electrocatalytic activity and outstanding conductivity for biomolecules detection.

3.3.1 Metal Oxides

Metal oxides are one of the most common products obtained using

MOFs precursors. Under air atmosphere, MOFs treatment at a particular temperature triggers the decomposition of the frameworks, leaving metal ions converted into metal oxides. With further tailoring the calcination conditions, metal oxides with specific properties, such as nanoporous metal oxides [51], can be synthesized. Due to the high electrocatalytic activity of transition metals toward glucose in the alkaline media [52], different metal oxides derived from MOFs are explored for the non-enzymatic glucose detection. Wang and co-workers proposed hierarchical NiO superstructures on foam Ni using Ni-MOFs on foam Ni as precursors [53]. The authors first prepared Ni-MOFs/foam Ni with a "twin metal source" method. Subsequently, NiO array/foam Ni electrode was obtained via calcination. The hierarchical NiO superstructures, uniformly embedded in three-dimensional carbon frameworks, effectively enhanced the electrical conductivity. The outstanding electrocatalytic activity of NiO array/foam Ni electrode toward glucose was demonstrated. Zhang and coworkers developed cobalt oxide hollow nanododecahedra (Co_3O_4-HND) with a facile thermal treatment by using ZIF-67 as precursor [54]. Results showed that Co_3O_4-HND inherited the dimensions and shape of ZIF-67 precursor with abundant pores formed during calcination. Co_3O_4-HND exhibited high activity and an outstanding performance for determination of glucose.

3.3.2 Metal/Carbon Nanocomposites

Loading metal nanomaterials on conducive carbon matrix, e.g., carbon nanotubes [55] or graphene [7], can be an effective way to boost the electroanalytical performance. The carbon matrix is able to enlarge the specific surface area and improve the loading stability of metal nanomaterials with enhanced electron conductivity. However, complicated synthetic steps are often employed in these chemical processes [56,57]. In 2012, Poddar *et al.* [58] systematically investigated the formation of metal/carbon nanohybrid. The authors found that that the MOFs comprising metal unities with a reduction potential equal to or higher than -0.27 V (such as copper and cobalt) resulted in metal/carbon nanocomposites by calcination. Guided by this rule, various metal/carbon nanocomposites can be obtained by using MOFs as precursors under inert atmosphere with an efficient one-step calcination process, and the resulting nanocomposites exhibit outstanding electrochemical performance. Moreover, it has been demonstrated that during the carbonization of organic ligands, the formed metal nanoparticles could act as catalysts, facilitating the transformation of organic linkers into graphite carbon [59,60] with enhanced electron conductivity.

Wei and co-workers synthesized 3D anthill-like Cu@carbon nano-composites by employing $Cu_3(BTC)_2$ as precursor, using a simple thermolysis [61]. The morphology of the pristine MOF was well re-tained after the thermo-treatment. The resulting Cu-C nanocompo-sites exhibited high electrocatalytic activity toward glucose. Recently, Shu and co-workers proposed Ni-MOF/Ni/NiO/C nanocomposite us-ing an efficient one-step calcination method by employing Ni-MOF as precursor (Figure 3.3) [62]. By calcination under argon, Ni, NiO and carbon frame were obtained simultaneously on Ni-MOF, as shown in Figure 3.3B. Ni-MOF/Ni/NiO/C showed high electrocatalytic activity toward glucose oxidation in alkaline media.

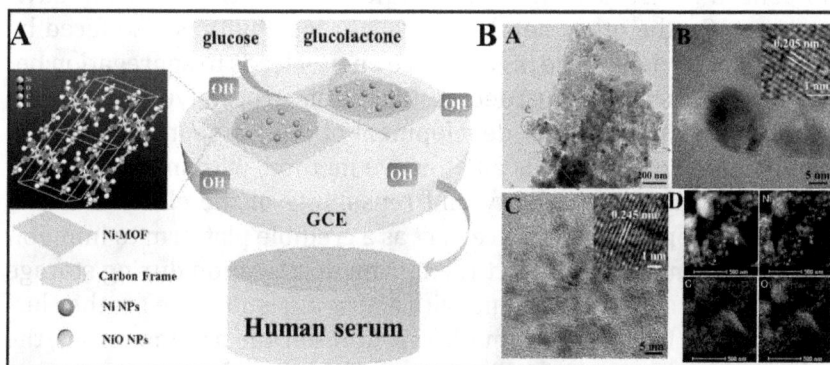

Figure 3.3 (A) Schematic of the Ni-MOF/Ni/NiO/C nanocomposite-modified electrode applied for glucose detection in human serum sample, (B) is the corresponding transmission electron microscopic characterization of Ni-MOF/Ni/NiO/C nanocomposite and the corresponding Ni and NiO nanoparticles. Reproduced from Reference 62 with permission from American Chemical Society.

As for the MOFs composed of metals with relatively low boiling point, such as zinc (904 °C), calcination treatment with higher tem-perature under nitrogen can generate a final product of porous car-bon without metals. In this procedure, the MOF precursors act as a self-sacrificial template and the obtained porous carbon can inherit the shape or morphology of the as-synthesized MOF with unexpect-edly high surface area and good electrochemical properties [63,64]. Gai and co-workers carbonized ZIF-8 under nitrogen to obtain nitrogen-doped porous carbon polyhedral (N-PCNPs) [65]. The developed N-PCNPs showed uniform polyhedral morphology and a high surface area of 2221 m^2g^{-1}. N-PCNPs also showed the capability of simultaneous detection of ascorbic acid, dopamine and uric acid in human urine sample.

3.4 MOFs as Immobilization Hosts

Although significant attention has been focused on exploring nano-materials-based electrocatalysts as an alternative strategy to natural enzymes, conventional enzymatic electroanalysis is still a reliable method due the higher reaction rate and selectivity compared with non-enzymatic process. The main issue faced by enzymatic electroanalysis is the fragile nature of enzymes, which makes the detection influenced by temperature, pH and humidity of the detection environment. On the other hand, the issue of instability is also faced by the nanomaterials as the high surface energy leads to aggregation between nanoparticles, thus, decreasing the electrocatalytic activity. To overcome these problems, development of solid support materials to immobilize the catalysts is an effective method, with improved electrocatalytic activity, stability and reusability of the electrocatalyst. The solid support materials can act as a credible platform to immobilize the enzymes and protect them from deactivation during storage and catalytic conditions, along with acting as a substrate for the electron transfer from the enzymatic reaction. For the nanomaterials, the solid supports can effectively control their size, shape and aggregation, for effective application in non-enzymatic electroanalysis.

Till now, various solid support materials, including conducting polymers [66,67], carbon [68,69] and other metallic nanocomposites [70,71], have been developed. Recently, MOFs have also become a promising candidate because of the high specific surface area, ease of tuning pore size and functionality and thermal as well as chemical stability, thus, providing an outstanding platform for the immobilization/encapsulation of bioactive enzymes or nanomaterials. Moreover, various synthesis strategies for MOFs-based nanocomposites have been investigated. A few mild synthesis strategies at room temperature allow the composites to retain their high activity. As for the immobilization strategy for the biocatalysts (e.g., enzymes and hemin), several procedures have also been studied, including surface attachment, covalent linkage, pore encapsulation, co-precipitation, etc. The mechanism and characteristic of these immobilization methods have also been reviewed in detail by Lian and co-workers [72]. Moon and co-workers have also reviewed the various methods proposed for the synthesis of MOFs-metal nanocomposites [73]. In addition, a synthetic overview of the synthesis strategies and application of diverse MOFs composites by Zhu and Xu [12] can also be referred to.

Herein, we discuss MOFs as potential immobilization hosts from the perspective of the immobilization composites, which are divided into biorecognition units and inorganic nanomaterials.

3.4.1 Biorecognition Units

Mao *et al.* [74] were the first to explore the application of zeolitic imidazolate frameworks (ZIF) as a matrix for electroanalysis. They synthesized various ZIF materials including ZIF-7, ZIF-8, ZIF-67, ZIF-68, ZIF-70 with various structural topologies, functional groups, surface areas, metal centers and pore sizes. Glucose dehydrogenase (GDH) as biorecognition units and methylene green (MG) as electron mediator were co-immobilized in ZIF materials. Among these ZIF materials, ZIF-70 exhibited the highest adsorption ability toward MG due to the surface area and d_a, d_g values, with the ability of maintaining the electrochemical activity of MG. The adsorption capability toward GDH of ZIF-70 was only lower than ZIF-68 because of the presence of strongly hydrophobic benzimidazole ligand in ZIF-68. Overall, the electrochemical sensor based on ZIF-70 as the immobilization matrix showed superior performance toward practical *in vivo* measurements on brain glucose of living guinea pigs.

Wang and coworkers applied 3D Cu-MOF as a matrix to immobilize enzymes for bisphenol A (BPA) electrochemical detection [75]. Tyrosinase (Tyr) was firstly mixed with Cu-MOF by simple shaking, followed by the modification of the mixture on the glassy carbon electrode (GCE) by chitosan (Chit). The sensitivity of Cu-MOF-Tyr-Chit/GCE toward BPA was proved to be 2 times that of the enzyme electrode without Cu-MOF. It was inferred that Cu-MOF was not only beneficial for the enzyme adsorption due to the large surface area, but also increased the available BPA concentration to react with tyrosinase because of the π-π stacking interactions with BPA. To improve the loading ability, Chen *et al.* [76] developed boronic acid functionalized metal organic frameworks (MIL-100(Cr)-B) for the effective immobilization of horseradish peroxidase (HRP), as shown in Figure 3.4A. The strong interaction between boronic acid and sugar can efficiently entrap HRP as the typical glycoprotein. With the hierarchical porous structure, extremely high surface area and sufficient recognition sites of MIL-100(Cr)-B, it increased HRP loading and prevented leakage as well as deactivation. The resulted HRP-MIL-100(Cr)-B also showed low cytotoxicity, and the production of H_2O_2 from RAW 264.7, Hela and MCF-7 living cells was detected. Results showed that the production of H_2O_2 from the cancer cells was more notable than the abnormal RAW 246.7 cells [76].

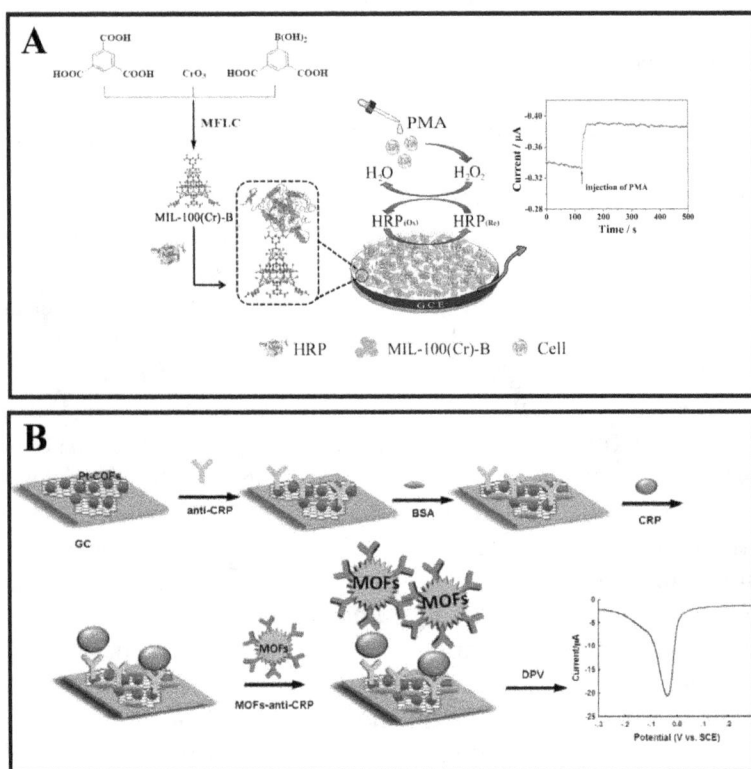

Figure 3.4 (A) Schematic illustration for the boronic acid functionalized MOFs (MIL-100(Cr)-B) with the immobilization of HRP for the electrochemical detection of H_2O_2 release from cells stimulated with Phorbol 12-myristate 13-acetate (PMA). Reproduced from Reference 76 with permission from Elsevier. (B) Schematic illustration for electrochemical immuosensor for the detection of C-reactive protein (CRP). Cu-MOF was used for the immobilization of CRP antibody, meanwhile providing an electrochemical signals from the Cu ions in the MOFs structure. Reproduced from Reference 77 with permission from American Chemical Society.

MOFs as immobilization hosts are also found to show the promising application in electrochemical immunosensors. In the sandwich-type electrochemical immunoassay, antibody can be efficiently loaded on MOFs as an alternative to the conventional gold nanoparticles. This mode also shows a potential of easier operation and higher response due to some MOFs hosts themselves having a sensitive electrochemical signal from the metal species in the MOFs structure. For example, Liu and coworkers used Cu-MOFs (also known as HKUST-1) to label the signal antibody (Ab2) to perform the sandwich-type immuno-detection of C-reactive protein (CRP), as shown in Figure 3.4B [77]. The large surface area of MOFs provided an outstanding platform to immobilize the CRP antibody forming the signal-MOFs-metal ion probes. Cu-MOF itself could act as a signal probe to quantify the target CRP due to the presence of Cu^{2+} in the structure, which showed a strong reduction peak in the electrochemical detection without acid dissolution and preconcentration. The method greatly simplified the detection steps and reduced the detection time.

Additionally, other studies have also confirmed MOFs to be outstanding platforms to immobilize biorecognition units, such as horseradish peroxidase [78], microperoxidase-11 [79,80], hemin [81], etc., for various electroanalysis applications.

3.4.2 Inorganic Nanomaterials

The high surface area and porous features also allow MOFs to become superior immobilization hosts for inorganic nanomaterials for non-enzymatic electroanalysis. In 2012, Hosseini and coworkers reported, for the first time, the application of MOFs supported metal nanoparticles for electroanalysis of L-cysteine (CySH) [82]. Thiol-modified SiO_2 (SH-SiO_2) was firstly immobilized in Cu-MOF. Au nanoparticles were subsequently linked on the composites forming Au-SH-SiO_2@Cu-MOF. Results showed the Au-SH-SiO_2@Cu-MOF composites had a higher oxidation current and lower overpotential toward the oxidation of CySH. The catalytic activity could be attributed to the high surface area and an increase in the rate of the electron transfer process. Fernandes and coworkers established a polyoxometalates (POM)@MOFs system for non-enzymatic electroanalysis [83,84]. For example, they encapsulated a kind of POM, $PMo_{10}V_2$, into the porous MIL-101(Cr) for the electrochemical detection of ascorbic acid (AA). Due to the porous structure of MIL-101, the MOF substrate allows the immobilization of larger quantities of electroactive $PMo_{10}V_2$. The results showed improved electrocatalytic efficiency toward AA on $PMo_{10}V_2$@MIL-101 than without MOFs host. Recently, our group applied zeolite imidazole framework-8 (ZIF-8) as an immobilization host for encapsulating Cu nanoparticles (CuNPs) for non-enzymatic glucose sensing [85]. The nanostructure was synthesized with an impregnation-reduction method. The method allowed the growth of Cu nanoparticles inside the cage of ZIF-8. Due to the steric restriction effect of these cages, the size of CuNPs was confined in the range from 2 to 5 nm, which remarkably increased the activity of the electrocatalyst. Moreover, the cages also protected CuNPs from electrochemical dissolution, migration and agglomeration during the electrocatalytic process, obtaining high stability with around 90% of initial response after 35 glucose detection cycles.

3.5 Conclusion and Prospects

In summary, a review of the various applications of MOFs in electrochemical sensors for the detection of biological molecules has been presented in this chapter. In the healthcare sector, the significance of biomolecule detection has led to an extensive research effort, especially for the electrochemical sensing. Improving the detection performance of sensors has also been a topic of interest due to its potential for obtaining the spatial and temporal information of the biomolecules in biological systems. Developing novel nanomaterials to modify the electrodes provides one of the most promising research directions for the high performance electrochemical sensors. Recently, MOFs have emerged as effective porous materials in this field. Based on the recent literature, the application of MOFs as the electrode-modifying materials has been largely successful owing to the distinguishing features of MOFs materials. To date, the application of MOFs in electrochemical sensors has been mainly divided into three aspects: (1) being utilized as the direct electrocatalysts owing to the electrocatalytic activity of the metallic species in MOFs; (2) being utilized as the calcination precursors for preparation of porous carbon, metal (or metal oxides) and their composites as electrocatalysts; (3) being utilized as the immobilization hosts for both inorganic nanomaterials and biorecognition units with high activity and stability. The characteristics of these applications including some representative examples have been well documented in this chapter.

It should be pointed out that the research on the application of MOFs in electrochemical sensors is still at its early stage. MOFs show significant potential in meeting the requirements of highly sensitive and selective detection sensors. Regarding these aspects, as also reviewed in this chapter, there are still significant areas requiring further work. For instance:

(1) For the application of MOFs as direct electrocatalysts, the hydrostability of MOFs is one of the key characteristics that need to be improved, considering most of the detection is performed in aqueous systems. Although some strategies handling the inert properties of the organic ligands have been discussed, these methods are either complicated or not applicable for all types of MOFs. In order to achieve optimal performance, it is still essential to develop novel strategies to increase the electrocatalytic activity of MOFs electrocatalysts.

(2) Calcination of MOFs is an efficient method to synthesize carbon and metallic nanocomposites. Based on the findings of Poddar *et al.* [58], we can conclude that the type of metal is highly restricted to its reduction potential. Thus, more focus needs to be paid be paid on exploring different calcination conditions, through which we can gain specific physical and chemical properties with increased electrocatalytic activity of the MOFs-derived composites. Additionally, either doping other metallic species or combination of MOFs with other materials as precursors followed by the calcination treatment will give us new directions to synthesize nanocomposites with desired properties and composition to construct the electrochemical sensors. This strategy could greatly expand application of MOFs-derived nanomaterials for the detection of biomolecules.

(3) For the application of MOFs as immobilization hosts, one of the most challenging issues is the synthesis method, especially for the immobilization of biorecognition units like enzymes onto MOFs. The synthesis conditions of MOFs-enzymes composites must be mild to prevent the denaturing of enzymes. For the electroactive materials immobilized in the pore structures of MOFs, while the pore structures would provide enhance the stability of the active materials, the hindrance effect of MOFs shell should also be taken into consideration. Thus, the design of the MOFs-active materials is also very important for developing high performance electrochemical sensors.

Till now, efforts are being made for developing novel nanomaterials for the electrochemical sensors for the purpose of improving the detection performance. However, we must admit that it is tremendously challenging to achieve an integrated commercialized detection device, which is effective for various analysis targets. This is a comprehensive work in which various fields such as electrochemistry, materials science, biomedicine, information technology and mechanical engineering need to be integrated to achieve meaningful products in future.

Acknowledgements

This work was financially supported by the Science and Technology Commission of Shanghai Municipality (STCSM, No. 16520710800) and the Fundamental Research Funds for the Central Universities (No. 222201817022).

References

1. Zhao, L., Sun, L., and Chu, X. (2009) Chemiluminescence immunoassay. *TrAC Trends in Analytical Chemistry,* **28**(4), 404-415.
2. Wang, Z., Zong, S., Wu, L., Zhu, D., and Cui, Y. (2017) SERS-activated platforms for immunoassay: probes, encoding methods, and applications. *Chemical Reviews,* **117**(12), 7910-7963.
3. Chinen, A. B., Guan, C. M., Ferrer, J. R., Barnaby, S. N., Merkel, T. J., and Mirkin, C. A. (2015) Nanoparticle probes for the detection of cancer biomarkers, cells, and tissues by fluorescence. *Chemical Reviews,* **115**(19), 10530-10574.
4. Xiao, T., Wu, F., Hao, J., Zhang, M., Yu, P., and Mao, L. (2017) In vivo analysis with electrochemical sensors and biosensors. *Analytical Chemistry,* **89**(1), 300-313.
5. Zhu, C., Yang, G., Li, H., Du, D., and Lin, Y. (2015) Electrochemical sensors and biosensors based on nanomaterials and nanostructures. *Analytical Chemistry,* **87**(1), 230-249.
6. Lim, W. Q., and Gao, Z. (2015) Metal oxide nanoparticles in electroanalysis. *Electroanalysis,* **27**(9), 2074-2090.
7. Shixin, W., Qiyuan, H., Chaoliang, T., Yadong, W., and Hua, Z. (2013) Graphene-based electrochemical sensors. *Small,* **9**(8), 1160-1172.
8. Guo, S., and Wang, E. (2011) Noble metal nanomaterials: Controllable synthesis and application in fuel cells and analytical sensors. *Nano Today,* **6**(3), 240-264.
9. Jaiswal, N., and Tiwari, I. (2017) Recent build outs in electroanalytical biosensors based on carbon-nanomaterial modified screen printed electrode platforms. *Analytical Methods,* **9**(26), 3895-3907.
10. Farha, O. K., Yazaydın, A. O., Eryazici, I., Malliakas, C. D., Hauser, B. G., Kanatzidis, M. G., Nguyen, S. T., Snurr, R. Q., and Hupp, J. T. (2010) *De*

novo synthesis of a metal-organic framework material featuring ultrahigh surface area and gas storage capacities. *Nature Chemistry,* **2**(11), 944-948.

11. Farha, O. K., Eryazici, I., Jeong, N. C., Hauser, B. G., Wilmer, C. E., Sarjeant, A. A., Snurr, R. Q., Nguyen, S. T., Yazaydın, A. O., and Hupp, J. T. (2012) Metal–organic framework materials with ultrahigh surface areas: Is the sky the limit? *Journal of the American Chemical Society,* **134**(36), 15016-15021.

12. Zhu, Q. L. and Xu, Q. (2014) Metal-organic framework composites. *Chemical Society Reviews,* **43**(16), 5468-5512.

13. Murray, L. J., Dinca, M., and Long, J. R. (2009) Hydrogen storage in metal-organic frameworks. *Chemical Society Reviews,* **38**(5), 1294-1314.

14. Li, J. R., Kuppler, R. J., and Zhou, H. C. (2009) Selective gas adsorption and separation in metal-organic frameworks. *Chemical Society Reviews,* **38**(5), 1477-1504.

15. Zhuang, J., Kuo, C- H., Chou, L.-Y., Liu, D.-Y., Weerapana, E., and Tsung, C.-K. (2014) Optimized metal–organic-framework nanospheres for drug delivery: evaluation of small-molecule encapsulation. *ACS Nano,* **8**(3), 2812-2819.

16. Wuttke, S., Lismont, M., Escudero, A., Rungtaweevoranit, B., Parak, W. J. (2017) Positioning metal-organic framework nanoparticles within the context of drug delivery - A comparison with mesoporous silica nanoparticles and dendrimers. *Biomaterials,* **123**, 172-183.

17. Gao, X., Zhai, M., Guan, W., Liu, J., Liu, Z., and Damirin A. (2017) Controllable synthesis of a smart multifunctional nanoscale metal–organic framework for magnetic resonance/optical imaging and targeted drug delivery. *ACS Applied Materials and Interfaces,* **9**(4), 3455-3462.

18. Rossin, A., Tuci, G., Luconi, L., and Giambastiani, G. (2017) Metal-organic frameworks as heterogeneous catalysts in hydrogen production from lightweight inorganic hydrides. *ACS Catalysis,* **7**(8), 5035-5045.

19. Zhu, L., Liu, X. Q., Jiang, H. L., and Sun, L. B. (2017) Metal–organic frameworks for heterogeneous basic catalysis. *Chemical Reviews,* **117**(12), 8129-8176.

20. Wu, C. D., and Zhao, M. (2017) Incorporation of molecular catalysts in metal-organic frameworks for highly efficient heterogeneous catalysis. *Advanced Materials,* **29**(14), 1605446.

21. Clark, L. C., and Lyons, C. (1962) Electrode systems for continuous monitoring in cardiovascular surgery. *Annals of the New York Academy of Sciences,* **102**(1), 29-45.

22. Park, S., Boo, H., and Chung, T. D. (2006) Electrochemical non-enzymatic glucose sensors. *Analytica Chimica Acta,* **556**(1), 46-57.

23. Ju, J., and Chen, W. (2015) In situ growth of surfactant-free gold

nanoparticles on nitrogen-doped graphene quantum dots for electrochemical detection of hydrogen peroxide in biological environments. *Analytical Chemistry,* **87**(3), 1903-1910.

24. Maji, S. K., Sreejith, S., Mandal, A. K., Ma, X., and Zhao, Y. (2014) Immobilizing gold nanoparticles in mesoporous silica covered reduced graphene oxide: A hybrid material for cancer cell detection through hydrogen peroxide sensing. *ACS Applied Materials & Interfaces,* **6**(16), 13648-13656.

25. Yin, J., Qi, X., Yang, L., Hao, G., Li, J., and Zhong, J. (2011) A hydrogen peroxide electrochemical sensor based on silver nanoparticles decorated silicon nanowire arrays. *Electrochimica Acta,* **56**(11), 3884-3889.

26. Niu, X., Shi, L., Pan, J., Qiu, F., Yan, Y., Zhao, H., and Lan, M. (2016) Modulating the assembly of sputtered silver nanoparticles on screen-printed carbon electrodes for hydrogen peroxide electroreduction: effect of the surface coverage. *Electrochimica Acta,* **199**, 187-193.

27. Li, X., Liu, X., Wang, W., Li, L., and Lu, X, (2014) High loading Pt nanoparticles on functionalization of carbon nanotubes for fabricating nonenzyme hydrogen peroxide sensor. *Biosensors and Bioelectronics,* **59**, 221-226.

28. Liu, J., Bo, X., Zhao, Z., and Guo, L. (2015) Highly exposed Pt nanoparticles supported on porous graphene for electrochemical detection of hydrogen peroxide in living cells. *Biosensors and Bioelectronics,* **74**, 71-77.

29. Liu, M., Liu, R., and Chen, W. (2013) Graphene wrapped Cu_2O nanocubes: Non-enzymatic electrochemical sensors for the detection of glucose and hydrogen peroxide with enhanced stability. *Biosensors and Bioelectronics,* **45**, 206-212.

30. Chirizzi, D., Guascito, M. R., Filippo, E., Malitesta, C., and Teporea, A. (2016) A novel nonenzymatic amperometric hydrogen peroxide sensor based on $CuO@Cu_2O$ nanowires embedded into poly(vinyl alcohol). *Talanta,* **147**, 124-131.

31. Qin, Z., Zhao, Y., Lin, L., Zou, P., Zhang, L., Chen, H., Wang, Y., Wang, G., and Zhang, Y. (2017) Core/shell microcapsules consisting of Fe_3O_4 microparticles coated with nitrogen-doped mesoporous carbon for voltammetric sensing of hydrogen peroxide. *Microchimica Acta,* **184**(11), 4513-4520.

32. Sun, X., Guo, S., Liu, Y., and Sun, S. (2012) Dumbbell-like PtPd–Fe_3O_4 nanoparticles for enhanced electrochemical detection of H_2O_2. *Nano Letters,* **12**(9), 4859-4863.

33. Wu, W., Li, Y., Jin, J., Wu, H., Wang, S., and Xia, Q. (2016) A novel nonenzymatic electrochemical sensor based on 3D flower-like Ni7S6 for hydrogen peroxide and glucose. *Sensors and Actuators B: Chemical,* **232**, 633-641.

34. Yan, Q., Wang, Z., Zhang, J., Peng, H., Chen, X., Hou, H., and Liu, C. (2012) Nickel hydroxide modified silicon nanowires electrode for hydrogen peroxide sensor applications. *Electrochimica Acta,* **61**, 148-153.

35. Li, B., Chen, D., Wang, J., Yan, Z., Jiang, L., Duan, D., He, J., Luo, Z., Zhang, J., and Yuan, F. (2014) MOFzyme: Intrinsic protease-like activity of Cu-MOF. *Scientific Reports,* **4**, 6759.

36. Liu, Y., Zhang, Y., Chen, J., and Pang, H. (2014) Copper metal-organic framework nanocrystal for plane effect nonenzymatic electro-catalytic activity of glucose. *Nanoscale,* **6**(19), 10989-10994.

37. Zhang, D., Shi, H., Zhang, R., Zhang, Z., Wang, N., Li, J., Yuan, B., Bai, H., and Zhang, J. (2015) Quick synthesis of zeolitic imidazolate framework microflowers with enhanced supercapacitor and electrocatalytic performances. *RSC Advances,* **5**(72), 58772-58776.

38. Yuan, B., Zhang, R., Jiao, X., Li, J., Shi, H., and Zhang, D. (2014) Amperometric determination of reduced glutathione with a new Co-based metal-organic coordination polymer modified electrode. *Electrochemistry Communications,* **40**, 92-95.

39. Wang, M. Q., Ye, C., Bao, S. J., Zhang, Y., Yu, Y. N., and Xu, M. (2016) Carbon nanotubes implanted manganese-based MOFs for simultaneous detection of biomolecules in body fluids. *Analyst,* **141**(4), 1279-1285.

40. Wang, M. Q., Zhang, Y., Bao, S. J., Yu, Y. N., and Ye, C. (2016) Ni(II)-Based metal-organic framework anchored on carbon nanotubes for highly sensitive non-enzymatic hydrogen Peroxide Sensing. *Electrochimica Acta,* **190**, 365-370.

41. Yang, Y., Wang, Q., Qiu, W., Guo, H., and Gao, F. (2016) Covalent immobilization of Cu3(btc)2 at chitosan-electroreduced graphene oxide hybrid film and its application for simultaneous detection of dihydroxybenzene isomers. *The Journal of Physical Chemistry C,* **120**(18), 9794-9803.

42. Wang, Q., Yang, Y., Gao, F., Ni, J., Zhang, Y., and Lin, Z. (2016) Graphene oxide directed one-step synthesis of flowerlike graphene@HKUST-1 for enzyme-free detection of hydrogen peroxide in biological samples. *ACS Applied Materials & Interfaces,* **8**(47), 32477-32487.

43. Chang, Z., Gao, N., Li, Y., and He, X. (2012) Preparation of ferrocene immobilized metal-organic-framework modified electrode for the determination of acetaminophen. *Analytical Methods,* **4**(12), 4037-4041.

44. Xu, Y., Yin, X. B., He, X. W., and Zhang, Y. K. (2015) Electrochemistry and electrochemiluminescence from a redox-active metal-organic framework. *Biosensors and Bioelectronics,* **68**, 197-203.

45. Zhang, D., Zhang, J., Shi, H., Guo, X., Guo, Y., Zhang, R., and Yuan, B. (2015) Redox-active microsized metal-organic framework for efficient nonenzymatic H_2O_2 sensing. *Sensors and Actuators B:*

Chemical, **221**, 224-229.

46. Shi, L., Niu, X., Zhao, H., and Lan, M. (2017) Significantly improved electrocatalytic activity of copper-based structures that evolve from a metal-organic framework induced by cathodization treatment. *ChemElectroChem,* **4**(2), 246-251.

47. Salunkhe, R. R., Kaneti, Y. V., Kim, J., Kim, J. H., and Yamauchi, Y. (2016) Nanoarchitectures for metal-organic framework-derived nanoporous carbons toward supercapacitor applications. *Accounts of Chemical Research,* **49**(12), 2796-2806.

48. Salunkhe, R. R., Kaneti, Y. V., and Yamauchi, Y. (2017) Metal–organic framework-derived nanoporous metal oxides toward supercapacitor applications: progress and prospects. *ACS Nano,* **11**(6), 5293-5308.

49. Mai, H. D., Rafiq, K., and Yoo, H. (2017) Nano metal-organic framework-derived inorganic hybrid nanomaterials: synthetic strategies and applications. *Chemistry - A European Journal,* **23**(24), 5631-5651.

50. Cao, X. H., Tan, C. L., Sindoro, M., and Zhang, H. (2017) Hybrid micro-/nano-structures derived from metal-organic frameworks: preparation and applications in energy storage and conversion. *Chemical Society Reviews,* **46**(10), 2660-2677.

51. Salunkhe, R. R., Tang, J., Kamachi, Y., Nakato, T., Kim, J. H., and Yamauchi, Y. (2015) Asymmetric supercapacitors using 3D nanoporous carbon and cobalt oxide electrodes synthesized from a single metal–organic framework. *ACS Nano,* **9**(6), 6288-6296.

52. Toghill, K. E., and Compton, R. G. (2010) Electrochemical non-enzymatic glucose sensors: a perspective and an evaluation. *International Journal of Electrochemical Science,* **5**(9), 1246-1301.

53. Wang, L., Xie, Y., Wei, C., Lu, X., Li, X., and Song, Y. (2015) Hierarchical NiO superstructures/foam Ni electrode derived from Ni metal-organic framework flakes on foam Ni for glucose sensing. *Electrochimica Acta,* **174**, 846-852.

54. Zhang, E., Xie, Y., Ci, S., Jia, J., and Wen, Z. (2016) Porous Co_3O_4 hollow nanododecahedra for nonenzymatic glucose biosensor and biofuel cell. *Biosensors and Bioelectronics,* **81**, 46-53.

55. Jacobs, C. B., Peairs, M. J., and Venton, B. J. (2010) Review: Carbon nanotube based electrochemical sensors for biomolecules. *Analytica Chimica Acta,* **662**(2), 105-127.

56. Tan, C., Huang, X., and Zhang, H. (2013) Synthesis and applications of graphene-based noble metal nanostructures. *Materials Today,* **16**(1), 29-36.

57. Khan, M., Tahir, M. N., Adil, S. F., Khan, H. U., Siddiqui, M. R. H., Al-warthan, A. A., and Tremel, W. (2015) Graphene based metal and metal oxide nanocomposites: synthesis, properties and their applications. *Journal of Materials Chemistry A,* **3**(37), 18753-18808.

58. Das, R., Pachfule, P., Banerjee, R., and Poddar, P. (2012) Metal and metal oxide nanoparticle synthesis from metal organic frameworks (MOFs): finding the border of metal and metal oxides. *Nanoscale,* **4**(2), 591-599.

59. Tang, J., Salunkhe, R. R., Zhang, H., Malgras, V., Ahamad, T., Alshehri, S. M., Kobayashi, N., Tominaka, S., Ide, Y., Kim, J. H., and Yamauchi, Y. (2016) Bimetallic metal-organic frameworks for controlled catalytic graphitization of nanoporous carbons. *Scientific Reports,* **6**, 30295.

60. Torad, N. L., Hu, M., Ishihara, S., Sukegawa, H., Belik, A. A., Imura, M., Ariga, K., Sakka, Y., and Yamauchi, Y. (2014) Direct synthesis of MOF-derived nanoporous carbon with magnetic Co nanoparticles toward efficient water treatment. *Small,* **10**(10), 2096-2107.

61. Wei, C., Li, X., Xu, F., Tan, H., Li, Z., Sun, L., and Song, Y. (2014) Metal organic framework-derived anthill-like Cu@carbon nanocomposites for nonenzymatic glucose sensor. *Analytical Methods,* **6**(5), 1550-1557.

62. Shu, Y., Yan, Y., Chen, J., Xu, Q., Pang, H., and Hu, X. (2017) Ni and NiO nanoparticles decorated metal–organic framework nanosheets: facile synthesis and high-performance nonenzymatic glucose detection in human serum. *ACS Applied Materials & Interfaces,* **9**(27), 22342-22349.

63. Cui, L., Wu, J., and Ju, H. (2015) Synthesis of bismuth-nanoparticle-enriched nanoporous carbon on graphene for efficient electrochemical analysis of heavy-metal ions. *Chemistry – A European Journal,* **21**(32), 11525-11530.

64. Torad, N. L., Hu, M., Kamachi, Y., Takai, K., Imura, M., Masataka, I., Naitoa, M., and Yamauchi, Y. (2013) Facile synthesis of nanoporous carbons with controlled particle sizes by direct carbonization of monodispersed ZIF-8 crystals. *Chemical Communications,* **49**(25), 2521-2523.

65. Gai, P., Zhang, H., Zhang, Y., Liu, W., Zhu, G., Zhang X., and Chen, J. (2013) Simultaneous electrochemical detection of ascorbic acid, dopamine and uric acid based on nitrogen doped porous carbon nanopolyhedra. *Journal of Materials Chemistry B,* **1**(21), 2742-2749.

66. Al-Sagura, H., Komathia, S., Khan, M. A., Gurek, A. G., and Hassan, A. (2017) A novel glucose sensor using lutetium phthalocyanine as redox mediator in reduced graphene oxide conducting polymer multifunctional hydrogel. *Biosensors and Bioelectronics,* **92**, 638-645.

67. Zhang, L., Zhou, C., Luo, J., Long, Y., Wang, C., Yu, T., and Xiao, D. (2015) A polyaniline microtube platform for direct electron transfer of glucose oxidase and biosensing applications. *Journal of Materials Chemistry B,* **3**(6), 1116-1124.

68. Wooten, M., Karra, S., Zhang, M., and Gorski, W. (2014) On the direct electron transfer, sensing, and enzyme activity in the glucose

oxidase/carbon nanotubes system. *Analytical Chemistry*, **86**(1), 752-757.

69. Mani, V., Devadas, B., Chen, S. M. (2013) Direct electrochemistry of glucose oxidase at electrochemically reduced graphene oxide-multiwalled carbon nanotubes hybrid material modified electrode for glucose biosensor. *Biosensors and Bioelectronics*, **41**, 309-315.

70. Su, S., Sun, H., Xu, F., Yuwen, L., Fan, C., and Wang, L. (2014) Direct electrochemistry of glucose oxidase and a biosensor for glucose based on a glass carbon electrode modified with MoS_2 nanosheets decorated with gold nanoparticles. *Microchimica Acta*, **181**(13), 1497-1503.

71. Du Toit, H., and Di Lorenzo, M. (2014) Glucose oxidase directly immobilized onto highly porous gold electrodes for sensing and fuel cell applications. *Electrochimica Acta*, **138**, 86-92.

72. Lian, X., Fang, Y., Joseph, E., Wang, Q., Li, J., Banerjee, S., Lollar, C., Wanga, X., and Zhou, H. -C. (2017) Enzyme-MOF (metal-organic framework) composites. *Chemical Society Reviews*, **46**, 3386-3401.

73. Moon, H. R., Lim, D. W., and Suh, M. P. (2013) Fabrication of metal nanoparticles in metal-organic frameworks. *Chemical Society Reviews*, **42**(4), 1807-1824.

74. Ma, W., Jiang, Q., Yu, P., Yang, L., and Mao, L. (2013) Zeolitic imidazolate framework-based electrochemical biosensor for in vivo electrochemical measurements. *Analytical Chemistry*, **85**(15), 7550-7557.

75. Wang, X., Lu, X., Wu, L., and Chen, J. (2015) 3D metal-organic framework as highly efficient biosensing platform for ultrasensitive and rapid detection of bisphenol A. *Biosensors and Bioelectronics*, **65**, 295-301.

76. Dai, H., Lü, W., Zuo, X., Zhu, Q., Pan, C., Niu, X., Liu, J., Chen, H. L., and Chen, X. (2017) A novel biosensor based on boronic acid functionalized metal-organic frameworks for the determination of hydrogen peroxide released from living cells. *Biosensors and Bioelectronics*, **95**, 131-137.

77. Liu, T.-Z., Hu, R., Zhang, X., Zhang, K.-L., Liu, Y., Zhang, X.-B., Bai, R.-Y., Li, D., and Yang, Y.-H. (2016) Metal–organic framework nanomaterials as novel signal probes for electron transfer mediated ultrasensitive electrochemical immunoassay. *Analytical Chemistry*, **88**(24), 12516-12523.

78. Fan, Z., Wang, J., Nie, Y., Ren, L., Liu, B., and Liu, G. (2016) Metal-organic frameworks/graphene oxide composite: A new enzymatic immobilization carrier for hydrogen peroxide biosensors. *Journal of The Electrochemical Society*, **163**(3), B32-B37.

79. Gong, C., Chen, J., Shen, Y., Song, Y., Song, Y., and Wang, L. (2016) Microperoxidase-11/metal-organic framework/macroporous carbon for detecting hydrogen peroxide. *RSC Advances*, **6**(83), 79798-79804.

80. Gong, C., Shen, Y., Chen, J., Song, Y., Chen, S., Song, Y., and Wang, L. (2017) Microperoxidase-11@PCN-333 (Al)/three-dimensional macroporous carbon electrode for sensing hydrogen peroxide. *Sensors and Actuators B: Chemical,* **239,** 890-897.

81. Xie, S., Ye, J., Yuan, Y., Chai, Y., and Yuan, R. (2015) A multifunctional hemin@metal-organic framework and its application to construct an electrochemical aptasensor for thrombin detection. *Nanoscale,* **7**(43), 18232-18238.

82. Hosseini, H., Ahmar, H., Dehghani, A., Bagheri, A., Tadjarodi, A., and Fakhari, A. R. (2013) A novel electrochemical sensor based on metal-organic framework for electro-catalytic oxidation of L-cysteine. *Biosensors and Bioelectronics,* **42,** 426-429.

83. Fernande, D. M., Barbosa, A. D. S., Pires, J., Balula, S. S., Cunha-Silva, L., and Freire, C. (2013) Novel Composite Material Polyoxovanadate@MIL-101(Cr): A highly efficient electrocatalyst for ascorbic acid oxidation. *ACS Applied Materials & Interfaces,* **5**(24), 13382-13390.

84. Fernandes, D. M., Granadeiro, C. M., Paes de Sousa, P. M., Grazina, R., Moura, J. J. G., Silva, P., Almeida Paz, F. A., Cunha-Silva, L., Balula, S. S., and Freire, C. (2014) SiW11Fe@MIL-101(Cr) Composite: a novel and versatile electrocatalyst. *ChemElectroChem,* **1**(8), 1293-1300.

85. Shi, L., Zhu, X., Liu, T., Zhao, H., and Lan, M. (2016) Encapsulating Cu nanoparticles into metal-organic frameworks for nonenzymatic glucose sensing. *Sensors and Actuators B: Chemical,* **227,** 583-590.

4

PEBAX-MOF Membranes for CO₂ Separation

Sara Alkhoori, Muthukumaraswamy R. Vengatesan, Georgios Karanikolos and Vikas Mittal*,**

Department of Chemical Engineering, The Petroleum Institute (part of Khalifa University of Science and Technology), Abu Dhabi, UAE

Corresponding author: vik.mittal@gmail.com
**Current address*: Bletchington, Wellington County, Australia

4.1 Introduction

Currently, a significant research attention has been devoted to the development of polymeric membranes as an alternative to the conventional gas separation technologies. In this respect, efforts have been carried out to enhance the chemical and/or physical properties of various polymers. Polymeric membrane separation technology has meritorious properties, including low energy consumption, mechanical simplicity, easy scale up and smaller footprint [1,2]. However, polymeric membranes suffer from a trade-off between permeability and selectivity, i.e. polymers with high gas permeability generally have low gas selectivity and vice versa [2-4]. In order to overcome this issue, several inorganic fillers have been incorporated in the organic polymers to generate mixed matrix membranes (MMMs), which synergistically combine the easy processability of polymers with superior permeability and selectivity of inorganic materials. Recently, metal organic frameworks (MOFs) have been considered as outstanding candidates to be used as fillers for MMMs. Several MOFs, such as MIL-53, MIL-101, ZIF-7, ZIF-8, ZIF-90, UiO-66, UiO-67, CuBTC and CuTPA, have been specifically identified for use in gas separation membranes. Among these, HKUST-1 is a facile cost-effective MOF filler material due to its three-dimensional morphology with interconnected porous network, beneficial for effective gas separation.

Permeation of gases through the polymeric membranes is based on solution-diffusion mechanism [5-8]. Permeation of a gas through

Metal Organic Frameworks, edited by Vikas Mittal
© 2019 Central West Publishing, Australia

a membrane (P_e) depends on its solubility in the membrane (S) and its diffusion though the boundary layer of the membrane (D). The gas molecules travel from the feed side across the membrane thickness l to the permeate side based on the chemical potential gradient across the membrane (from high pressure feed gas side $P_{i,f}$ to lower pressure permeate side $P_{i,p}$). The rate of permeation or flux is given by the equation:

$$J = \frac{DS(P_{i,f} - P_{i,p})}{l}$$

The overall selectivity coefficient α_{12} is given by the equation:

$$\alpha_{12} = \frac{S_1 D_1}{S_2 D_2} = \frac{P_1}{P_2}$$

where 1 and 2 are the faster and slower components respectively. In the case of several component mixture, selectivity is given in terms of the slowest permeating gas for the sake of simplicity.

One of the arising applications of polymeric membranes is the separation of CO_2 from natural gas. A study on the effect of greenhouse gases mentioned that CO_2 level has increased by 30% since the pre-industrial time and if it continues at the same rate, it can lead to significant global warming. One way to solve this problem is by sequestering the CO_2 emissions [9]. Additionally, the separation of CO_2 from natural gas is very important for controlling corrosion during transport and distribution as well as catalyst poisoning. Previously, the most commonly used technology for CO_2 capture from natural gas has been amine treatment, however, its operational difficulties and challenges of keeping the solvent clean make polymeric membranes an attractive alternative. However, many challenges have to be overcome for the membrane technology to be economically viable. In gas processing, for instance, the main priorities are (1) to control hydrocarbon recovery and (2) to meet sales gas specifications. The minimum CO_2 content specification in the sales gas is < 2%. Since membranes cannot accomplish such high level of separation, multiple stages might be required. This would increase the complexity, energy consumption and cost. Comparing with amine unit in terms of hydrocarbon losses, Sridhar *et al.* [1] mentioned that for a single-stage membrane unit, the losses range from 2% to 10%, while the use of two-stage membrane unit decreases the loss to 2%-5%. Losses in the amine unit are less than 1% and are due to the entrainment of the hydrocarbons with flashed streams. A good membrane should have high selectivity and high permeability, however, there is a trade-off relationship between the two, which introduces another challenge. This can be seen by plotting $\log(\alpha_{ij})$ vs. $\log(P_i)$, and the performance of polymeric membranes usually falls within this trade-off of Robeson upper-bound. For the industry, materials that have separation properties above this limit are favored. In addition, it is important for the membranes to have high mass transfer area, fouling resistance and mechanical stability, along with reproducibility and low manufacturing costs. For this, a good understanding of the properties and gas transport behavior of the polymers is important [1].

Among the various polymer matrices used to develop membranes, thermoplastic elastomer PEBAX-4533 is widely used in commercial applications. It is a block copolymer composed of covalently bonded hard segments of polyamide (PA), which provide the mechanical support to the polymer, and soft segments of polyether (PE) with high free volume, which offer transport channels for gas permeation [5,6]. The general chemical structure of PEBAX is

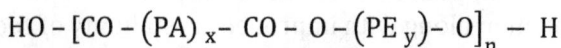

$$HO - [CO - (PA)_x - CO - O - (PE_y) - O]_n - H$$

Different PA:PE ratios result in different PEBAX grades. PEBAX polymers have wide range of applications as they combine properties of both thermoplastics and elastomers. One of the key factors affecting the performance of the membranes is the molecular structure, i.e. the arrangement of the chemical groups in the polymer chain. In the PEBAX polymer, PE units result in the amorphous phase which improves the permeability and the PA units lead to a crystalline phase which would enhance the stiffness, tensile strength and glass transition temperature (T_g). Thus, the ratio between the PE and PA blocks determines the performance of the MMM [10]. The covalent bonding between the PA and PE blocks prevents intermolecular mixing and a 'micro-phase separation' may arise. This is an important phenomenon of the block copolymers [3,4]. When micro-phase separation happens, the nanometer sized structures form uniquely arranged morphologies resulting in a micro-lattice structure. These arrangements can create micelles, spheres, cylinders, lamellae or surface patterning structure based on the relative length between the blocks.

Previous studies have reported different types of MMMs using different PEBAX grades containing altered PE and PA block lengths, along with a variety of nanofillers. A study by Murali *et al.* [3] investigated the effect of multi-walled carbon nanotubes (MWCNTs) (from 0 to 5 wt%) and crosslinking on the permeation behavior of PEBAX-1675 (60% PE and 40% PA). Incorporation of MWCNTs increased the intersegmental spacing of the polymer. Gas permeation analysis revealed that the filler improved the permeation of the polymer, specifically toward CO_2. Mixed matrix membranes of PEBAX-1657 loaded with 4A zeolite for gaseous separations were also developed by Murali *et al.* [6], and the main objective was to overcome the trade-off relationship limitations in polymeric membranes by utilizing the properties of organic polymer (good permeability) and inorganic zeolite (high selectivity due to small pore size and thermal stability). Other studies have also mentioned zeolites to be excellent fillers for separation purposes due to their molecular sieve ability and high thermal, mechanical and chemical stability [5]. A good dispersion of zeolite filler was observed for lower filler fractions. Also, corrugations were observed in the cross-sectional images which corresponded to the formation of void morphology due to the placement of the filler particles in between the polymer layers. Diffraction analysis exhibited an improvement in the amorphous nature (permeability) of the polymer, along with improved crystallinity of zeolite which was indicated by sharper peaks, with increasing zeolite loading. Polymer membranes filled with 10 wt% zeolite showed good permeability and selectivity. Moreover, permeability of CO_2 was observed to improve by increasing the pressure. Several literature studies have also carried out the permeability analysis of MMMs with HKUST-1 and most of them agreed that MOF enhanced the CO_2 separation performance of the polymeric membranes [11-14].

To further explore the usefulness of MOFs for gas separation membranes, MMMs involving PEBAX and HKUST-1 (in varying weight fractions) were developed and analyzed for their environmental and water stability as well as CO_2 permeation characteristics.

4.2 Experimental

4.2.1 Materials

PEBAX-4533 was kindly provided by ARKEMA in granular form. HKUST-1 MOF was synthesized at the Petroleum Institute laboratories by following the literature reported protocols. The synthesized HKUST-1 was a fine blue powder with a purity of 99%. The solvents used for the study were acquired from Sigma Aldrich.

4.2.2 Optimization of PEBAX-4533 Dissolution and Membrane Preparation

The dissolution of PEBAX-4533 was studied using different solvents such as N-methylpyrrolidone (NMP)/ethanol, dimethyl sulfoxide (DMSO) and dimethylformamide (DMF), along with optimum temperature and time. Around 10 mL of the solvent was used to dissolve 1 g of PEBAX-4533. In order to achieve a homogeneous solution, it was observed that the polymer granules should be stirred in the solvent at 80 °C for 4 h. It was observed that PEBAX dissolved uniformly in the most of the solvents. Overall, the PEBAX dissolution and membrane casting protocol was optimized with DMF as follows:

- Stir the polymer solution at 80 °C for 4 h.
- Pre-heat the casting dish before casting the solution. This step is to prevent the formation of lumps within the membrane film due to cooling down of the polymer solution while casting.
- Solvent evaporation in a pre-heated oven at 85 °C for 24 h.

4.2.3 Preparation of PEBAX-4533/HKUST-1 based MMMs

MMM formulations were prepared by stirring a specific amount of HKUST-1 in 1 g of PEBAX-4533 in DMF at 80 °C for 4 h. Similar casting procedure, as used for pure PEBAX membranes, was employed for generating MMMs. MMMs with different filler fractions of 0, 0.25, 0.5, 1 and 3 wt% were generated.

4.2.4 Characterization

The developed MMMs were tested for their water and environmental stability. For the environmental stability test, the samples were exposed directly to the outdoor environment without shade, while, for the water test, the samples were immersed in water in closed individual bottles at room temperature. In both tests, the samples were cut into 1 x 1 cm² size and were left for two weeks. The samples were characterized with thermogravimetric analysis (TGA), Fourier transform infra-red spectroscopy (FTIR) and scanning electron microscopy (SEM) before and after subjecting to environment and water conditions to study the (1) changes in the hydrogen bonding in the MMMs and (2) effect on the interfacial interactions between the MOF and polymer. TGA of the samples was carried out on a Discovery TGA from TA Instruments using nitrogen as carrier gas. 3-7 mg of the sample was heated from room temperature to 700 °C at a heating rate of 10 °C/min, and the weight loss as a function of temperature and time was recorded. For FTIR analysis, the samples were scanned between 4000-500 cm^{-1} with an accuracy of 0.27 cm^{-1} on a BRUKER Tensor II, Platinum ATR with mid-infrared range (MIR) detector at room temperature. For SEM, the membrane surface was analyzed in Hitachi field emission in-lens S-900 high resolution scanning electron microscope at accelerating voltages of 10-20 kV.

Permeability values were calculated using the steady state increase in the pressure with respect to time at a constant flow rate, as given by the equation:

$$P_i = \frac{J_i * l}{A * \left(P_{if} - P_{ip}\right)}$$

where P_i is the permeability coefficient of the component i in $\frac{cm^3(STP)cm}{s\ cm^2\ cmHg}$, J_i is the volumetric flow rate of permeated component in $\frac{cm^3}{s}$, l is the thickness of the membrane selective layer in cm, A is the effective membrane area per gas permeation in cm^2, P_{if} is the partial pressure of the component i at feed side and P_{ip} is the partial pressure of the component i at permeate side in $cmHg$.

Gases permeate through the membrane by solution-diffusion mechanism. At low pressures, CO_2 is mainly adsorbed on the metal centers of the MOF first followed by filling the small cages around the metal and organic linkers. In contrast, at higher pressures, adsorption is mainly in the vicinity of the rings and windows linking the pores [11-13].

Pure gas permeability analysis was carried out at room temperature with high-pressure gas permeation manifold built in the laboratory with nut and ferrule fittings for leak-free transportation of the gas streams. Figures 4.1 and 4.2 represent the in-house developed permeation manifold. The flow was controlled through needle valves with $\frac{1}{4}$ inch stainless steel 4BKT end connections. Before running the test, all lines along with the separation cell were flushed with pure CO_2. Pure CO_2 was used in all experiments at feed pressures of $P_{if} =$

3, 5 and 10 bar, while the permeate side pressure P_{ip} was kept at vacuum to deliver a driving force for the gas flow. MMM films were inserted in-between the stainless steel separation cell with an effective surface area A = 0.048 cm². Subsequently, the feed gas was allowed to follow through the separation cell and into the permeate side. After reaching a steady state, the data from the permeate side were collected through the Data Acquisition-Agilent 34972 and were transferred to the Benchlink software for producing a plot of pressure with respect to time.

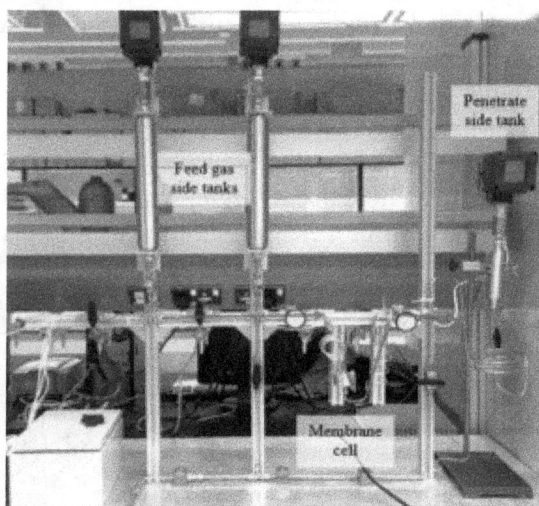

Figure 4.1 Gas permeation rig as well as feed gas and penetrate tanks.

Figure 4.2 a) Gas permeation rig, b) CO₂ feed and c) pump used for the permeation analysis.

4.3 Results and Discussion

From Figure 4.3 and Table 4.1, it can be seen that for the environmental stability test, the initial degradation of pure PEBAX as well as 0.25 wt% and 0.5 wt% composite samples occurred at significantly lower temperatures as compared to the original samples. This can be due to the harsh environment of high humidity and sunlight. However, for the membranes with filler fraction of 1 wt% and 3 wt%, the initial degradation was shifted to higher temperatures. The membranes were slightly hydrophilic in nature and underwent intermolecular interaction with the atmospheric moisture. For both environmental and water stability, the pure PEBAX membrane and 3 wt% MOF filled composite showed higher char yield due to the formation of H-bonded complex network. However, the lower filler fraction loaded composites exhibited uneven char yields. The higher filler loading also accelerated stronger hydrogen bonding between the MOF and the amide groups in PEBAX, resulting in improved thermal stability and char yield. The H-bonding in the nanocomposite was further confirmed by the FTIR analysis.

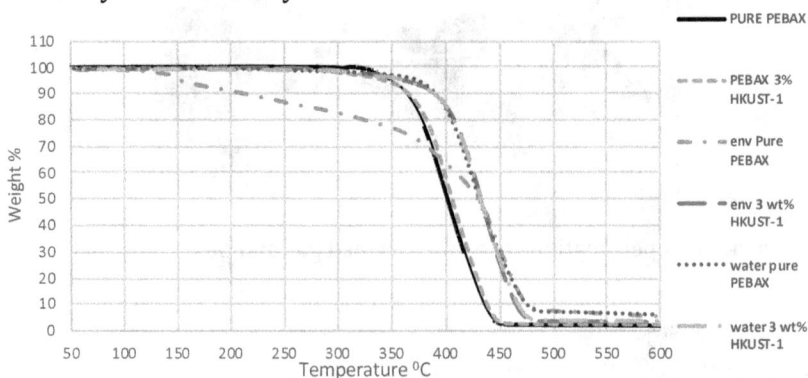

Figure 4.3 TGA plots of the MMMs before and after the environmental and water stability tests.

Hydrogen bonding plays an important role for achieving the stability in PEBAX based membranes. It has been reported that the addition of fillers can affect the hydrogen bonding state in the polymer as a result of the interfacial interaction between the MOF and polymer matrix, in addition to the effect of MOF on the polymer chain alignment and mobility [15,16]. More stable hydrogen bonding is observed to result in effective thermal stability of the MMMs. It has also been reported that the hydrogen bonding improves the structural order of crystalline polymers. PEBAX has highly polar carbonyl and amide groups that attribute to the hydrogen bonding in the polymer. Hydrogen bonds are the dominant interfaces holding the neighboring polymer chains together resulting in highly crystalline structure.

Table 4.1 TGA results of MMMs before and after environmental and water stability tests

Stability test	HKUST-1 wt%	5% degradation (°C)	10% degradation (°C)	Char yield (wt%)
Original samples	0	351	363	1.2
	0.25	353	366	1.6
	0.5	354	368	2.2
	1	354	367	2.5
	3	343	366	2.3
Environmental test	0	153	214	5.4
	0.25	19	24	-
	0.5	178	255	2.2
	1	373	393	1.1
	3	362	390	3.3
Water test	0	375	391	6.1
	0.25	375	393	-
	0.5	382	396	1.6-
	1	373	389	-
	3	365	389	3.7

FTIR spectra of MMMs before the stability tests showed that the free amide vibrations around 1734 cm⁻¹ remained unchanged, which indicated that the addition of MOF did not disturb the H-bonding of the amide groups. FTIR spectroscopy was subsequently carried out for the stability test samples to study the changes in hydrogen bonding. There are three bands for amide stretching in PEBAX: (i) hydrogen bonded carbonyl groups which are ordered crystalline domain, (ii) hydrogen bonded disordered amorphous conformation and (iii) non-hydrogen bonded free carbonyl groups. Therefore, the behavior of both amorphous and crystalline phases of PEBAX should be considered separately. Basically, if the amide characteristic bands are shifted to higher wavenumber or/and lower intensity, it indicates the disassociation of hydrogen bonding. This can occur only if the samples are exposed to higher temperatures such as in the case of environmental test. Moreover, the presence of peaks around 1680 cm⁻¹

and 1650 cm^{-1} is an indication of free and disordered hydrogen bonds of the carbonyl groups in the polymer. Figure 4.4 shows the FTIR spectroscopy of the water stability test samples plotted against the pure PEBAX membrane before the stability test.

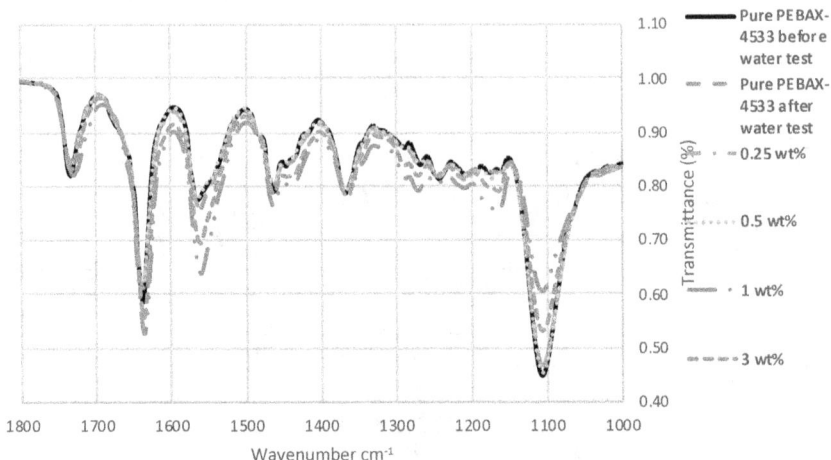

Figure 4.4 FTIR spectroscopy analysis of the water stability test samples and the pure PEBAX membrane before the stability test.

PEBAX characteristic peak around 1640 cm^{-1}, corresponding to the hydrogen-bonded amide groups, was slightly shifted to lower wavenumbers and increased in intensity for membranes with filler fractions of 1 and 3 wt%, indicating higher extent of hydrogen bonding. The other samples exhibited slightly lower intensity and also marginally shifted to lower wavenumbers.

For the free amide vibrations of $-C = O$, originally observed at 1734 cm^{-1}, the peak was observed to be broader with higher intensity for 1 wt% filler fraction along with a shift to lower wavenumber, thus, indicating the reconstruction of hydrogen bonding that hindered the chain mobility in the amorphous conformations, thereby, enhancing the thermal stability. However, the peaks for the rest of the samples were shifted slightly to lower wavenumbers and had decreased intensities.

Additionally, the FTIR analysis of two-week environmental stability test samples (pure PEBAX and composites with 0.5 and 1 wt% filler fraction) was carried out, as shown in Figure 4.5. The amide characteristic peaks around 1640 cm^{-1} for both composite membranes were slightly broadened and decreased in absolute intensity with no visible shift. These changes in the peaks demonstrated that the overall thermal stability and char yield of the membranes reduced after the environmental test.

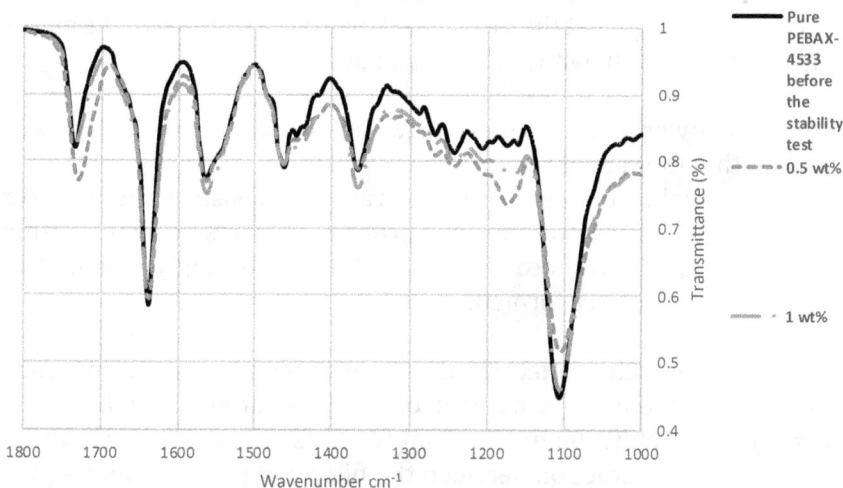

Figure 4.5 FTIR spectroscopy of the environmental stability test samples and the pure PEBAX membrane before the stability test.

For the free amide vibrations at 1734 cm^{-1}, peaks of both composite membranes were slightly broadened and shifted to lower wavenumbers. The peak corresponding to the 0.5 wt% membrane increased in absolute intensity suggesting the development of hydrogen bonding, while the intensity of the peak corresponding to 1 wt% membrane was slightly decreased.

Considering the hydrogen bonding in the crystalline domains of the polymer, it has been mentioned that the higher temperatures cause the crystalline lattice to expand, thus, increasing the distance between the chains. Therefore, the crystalline structure falls apart causing the ordered hydrogen bonds to turn into disordered bonds, thereby, decreasing the thermal stability. Hydrogen bonds in the crystalline domains were enhanced by effective interfacial interaction between the MOF and polymer matrix which was also revealed by SEM analysis, discussed later in this section. It was also shown by FTIR analysis that the interaction between the MOF and polymer was only physical, thus, the MOF acted like a physical crosslinking between the polymer chains, thereby, hindering the chain mobility and consequentially enhancing the thermal stability.

On the other hand, in the case of disordered amorphous domains in the polymers, increasing the temperature causes thermal expansion that weakens the strength of disordered hydrogen bonds. At highly elevated temperatures, these bonds may start to disassociate. It should be noted that the disordered hydrogen bonding in the amorphous phase is a dynamic equilibrium at which

> ➤ at lower temperatures, the rate of dissociation is much slower that the rate of construction
> ➤ at higher temperatures, the rate of reconstruction is faster since it is an endothermic process, thus, resulting in shifting the equilibrium to the dissociation. Accordingly, free hydrogen bonds are formed.

Also, other factors affect the direction shift of the equilibrium such as the presence of large number of highly polar groups (N − H and C = O) which shift the equilibrium towards reconstruction. Moreover, effective interaction between the filler and polymer matrix provides more surface area for the MOF to form hydrogen bonds with the amide groups in PEBAX.

The microstructure of MOF and MMMs was studied using SEM. From the SEM image of pure MOF (Figure 4.6), it can be seen that HKUST-1 exhibited a distinctive crystalline structure as octahedral shaped particles, and some of the particles are elongated towards a hexagonal symmetry [17]. The HKUST-1 crystals were also observed to aggregate with irregular morphology and size.

Figure 4.6 SEM micrograph of HKUST-1.

SEM images of the surface morphology of the pure PEBAX exhibited homogeneous and clear surface with no defects, cracks or plastic deformation (Figure 4.7a). On the other hand, the surface morphology of the MMMs (Figures 4.7b,c) exhibited the presence of sub-micron particles in the polymer matrix. The images showed that HKUST-1 particles were uniformly dispersed throughout the membrane without significant aggregations or defects. At higher magnifications, the polymer chains were observed to be wrapped around the filler crystals, suggesting a good interfacial interaction and adhesion between the phases.

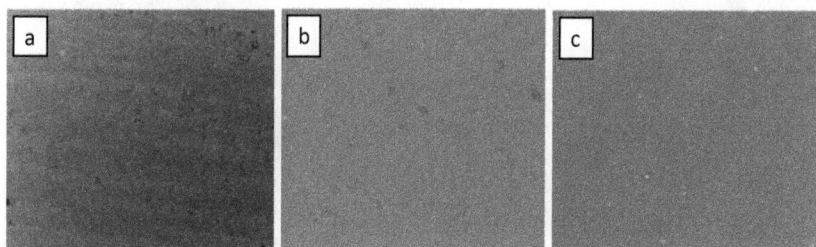

Figure 4.7 SEM surface images of a) pure PEBAX, b) 0.25 wt% MOF/PEBAX and c) 3.0 wt% MOF/PEBAX. The image width corresponds to 500 μm.

In the cross-sectional image analysis, PEBAX revealed a smooth horizontal layer morphology (Figure 4.8a). The morphology of the MMMs at lower filler fractions (Figure 4.8b) showed homogeneously distributed MOF within the polymer matrix. Even at higher loadings of MOF (Figure 4.8c), no significant cluster formation and aggregation were observed. In addition, no filler pull out was observed from the composite surface.

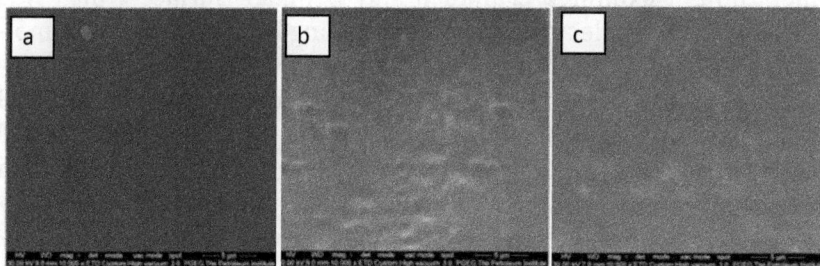

Figure 4.8 SEM cross-sectional images of a) pure PEBAX, b) 0.25 wt% MOF/PEBAX and c) 3.0 wt% MOF/PEBAX.

SEM analysis was also carried out for the water and environmental stability test samples in order to analyze the effect of nanoscale spatially constraining effect of MOF on the polymer chains. In the SEM images of the environmental stability test samples (0.5 wt% and 1 wt%) in Figure 4.9, MOF particles were observed to be randomly dispersed within the polymer matrix producing micro-void structure. As mentioned earlier, the effective interaction of the MOF particles with the polymer created physical crosslinking owing to the hydrogen bonding between the MOF and polymer chains, thus, hindering the chain mobility. The interaction was more pronounced in the 1 wt% membrane, which was also in good agreement with the TGA and FTIR results.

Figure 4.9 SEM micrographs of environmental stability test samples: a) 0.5 wt% and b) 1 wt% membranes.

Figure 4.10 represents the SEM images of the water stability test samples. The surface of the pure PEBAX membrane exhibited a rough morphology. The nanocomposite membranes with both low and high filler content exhibited crumbled and wrinkled surface which was due to the intermolecular hydrogen bonding between the water molecules, PEBAX and MOF. SEM micrographs exhibited an effective interfacial interaction between the MOF and polymer matrix for the 3 wt% membrane, which was in good agreement with the high thermal stability observed in TGA and induced hydrogen bond reconstruction observed in FTIR. Membrane with 0.25 wt% filler fraction also exhibited enhanced properties, however, at a lower magnitude as compared to other samples. This could be due to the lower filler loading, thus, lesser number of polar groups, which was not optimal to increase the reconstruction rate of the hydrogen bonds.

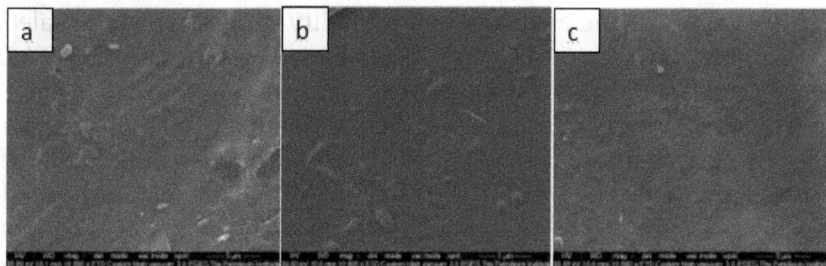

Figure 4.10 SEM micrographs of water stability test samples: a) pure PEBAX membrane, b) 0.25 wt% membrane and c) 3 wt% membrane.

PEBAX is a rubbery polymer, and the gas separation process is mainly through solution-diffusion mechanism since the volume fraction of the polymer matrix is much larger than that of HKUST-1 [5,12,13]. Permeability comprises of solubility, related to the compatibility of the filler and polymer matrix, and diffusivity through the membrane thickness [5]. Assuming good compatibility, diffusivity has a higher impact on the permeability as compared to solubility [5,11]. Incorporation of MOF enhances the CO_2 sorption and diffusion, along with delaying the plasticization owing to HKUST-1 characteristics. Diffusivity is highly dependent on the penetrate kinetic diameter, with CO_2 having a relatively small kinetic diameter of 3.3 Å, thus, leading to high solubility in PEBAX [13].

It should be noted that the permeation occurs only through the amorphous phase of the polymeric membrane. From the XRD analysis, it was confirmed that the pure PEBAX had amorphous and as well as crystalline phases. The incorporation of MOF did not exhibit any change in the crystalline phase in the composite, however, the flexible organic ligands of the filler may have interacted physically with the blocks of the copolymer, thus, resulting in an increase in the *d*-spacing and amorphous region of PEBAX. This indicated that the filler owing to its uniform distribution in PEBAX enhanced the permeability.

Moreover, CO_2 has high affinity towards MOF due to high quadrupole moment $(-1.4 * 10^{-35}$ C m) of CO_2, consequently resulting in the induction of molecular interactions between CO_2 and unsaturated Cu sites in the MOF, leading to the formation of H-bonds [11,14,18]. MOF increases the solubility of the condensable and polar gases in the polymer matrix by increasing the number of $-OH$ polar groups and altering the morphology at the interface between the permeate gas and polymer. This was further evidenced by the SEM and FTIR analysis of water and environmental stability studies, which revealed that a good interaction between the polymer matrix and HKUST-1 was due to the enhanced H-bonding.

Figure 4.11 and Table 4.2 show the permeation test results for the membranes at different CO_2 feed gas pressures of 3, 5 and 10 bar at 22 °C. First, in order to study the effect of MOF loading and the feed gas pressure on CO_2 permeability, the true permeability coefficients of the MMMs were plotted against the varying feed gas pressures. The pure PEBAX membrane exhibited the least permeability values as compared to the MOF filled matrix membranes. The permeability values of the MMMs were corrected with respect to MOF content. The membranes with 0.25 and 0.5 wt% filler fractions exhibited higher permeability than the pure PEBAX. This was attributed to the hindrance to the PEBAX chain mobility and possibility of the creation

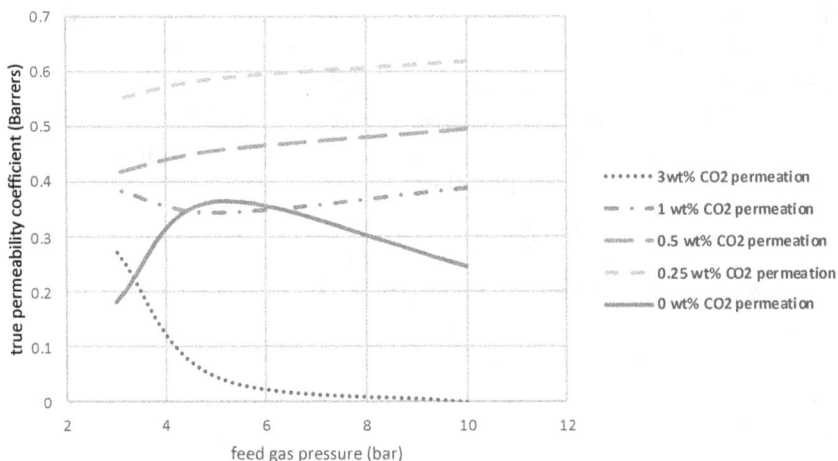

Figure 4.11 Effect of HKUST-1 loading and CO_2 feed gas pressure on permeation.

Table 4.2 True permeability coefficients (Barrer) of MMMs at different CO_2 feed gas pressures

pressure(bar)	0 wt%	0.25 wt%	0.5 wt%	1 wt%	3 wt%
10	0.246	0.619	0.494	0.39	0.168
5	0.364	0.587	0.457	0.344	0.0468
3	0.181	0.549	0.418	0.388	0.276

of voids at the interface. The membrane with 3.0 wt% HUKST-1 possessed lower permeability due to the filler aggregation at the microscale. Thus, the higher filler loadings blocked the pores available for permeation. HKUST-1 filled MMMs, as compared with other types of fillers, have higher possibility of decreasing the permeability due to the dominant molecular sieving behavior of HKUST-1. Furthermore, the penetration of the polymer chains may narrow the pore openings, thus, resulting in further reduction in diffusivity. Due to the flexible nature of the MOF organic ligands, these introduce the breathing effect, thus, improving the permeability, but, this can affect the separation performance negatively by limiting the molecular sieving phenomena [5,6,13,14]. Results from literature studies have shown that the breathing effect allows the HKUST-1 framework to contract and expand, consequently reducing or increasing the pore dimensions [5,11]. The enhancement in permeability indicates the absence of non-selective voids and good interaction between the filler and polymer matrix. It also suggests that composite membranes incorporated with 0.25 and 0.5 wt% HKUST-1 may be good candidates for CO_2 separation.

Secondly, in order to study the effect of CO_2 feed gas pressure, the permeability of each MMMs has been varied with feed gas pressure. In the case of 0.25 and 0.5 wt% MOF filled membranes, the permeability increased gradually with increasing the feed gas pressure and showed the highest values over the whole pressure range. Similar behavior has been reported in literature with PEBAX-1675 [13]. 1 wt% HKUST-1 filled membrane showed a 'dual-mode' sorption behavior which is also similar to the reported results [8,11]. A drop in permeability occurred at 5 bar pressure, which was attributed to the plasticization pressure level. The membranes with 0.25, 0.5 and 1.0 wt% MOF exhibited effective permeability at higher feed gas pressure of 10 bar of CO_2. At low feed gas pressure, CO_2 adsorption was preferentially around Cu sites followed by the vicinity around the metal center and organic ligands which resulted in a reduction in the solubility. On the other hand, at higher pressures, CO_2 diffused faster and filled the vicinity of the areas connecting the pores. It can be seen that the membrane with 3 wt% of HKUST-1 showed a significant decrease in permeability values throughout the whole pressure range probably owing to the aggregation of the filler particles, as mentioned earlier. It should also be noted that the trend at 10 bar pressure could be due to the non-ideal conditions.

As no experimental studies on PEBAX-4533/HKUST-1 MMMs

have been reported in the literature, thus, the comparisons were made with the nanocomposite membranes generated using other PEBAX grades. Table 4.3 summarizes the permeability coefficients reported in literature for PEBAX-1675 based MMMs with different fillers at similar experimental conditions of pressure and temperature.

Table 4.3 Permeability values of PEBAX-1675 nanocomposites

Filler	wt%	P (bar)	T (°C)	CO_2 permeability (Barrer)	Ref.
MWNT	0	10	30	55.9	3
	2			329	
	5			262	
	0	30		183.8	
Nano-Silica	0	10	30	1.1	14
	0.05			1.9	
	0.1			3.2	
	0.2			5.4	
	0.3			8.9	
H-Mordenite	0.05			1.6	
	0.1			2.1	
	0.2			3	
	0.3			5.4	
ZIF-8	2, 4, 8, 16	2 8		464 758	19
ZIF-8	0	4	25	55	20
	11			76	
	18			178	
	21			136	
	33			84	
4A zeolite	0	5	30	55	21
	5			71	
	10			97	
	20			113	
	30			155	
	10	25		130	
A4 zeolite	0	5	30	55.8	12
	5			71.4	
	10			97	
	20			113.7	
	30			155.8	

Pure PEBAX-1675 membranes have permeability coefficients around 55 Barrer at pressures ranging from 2 to 10 bar [3,12,14,19-

21], while the values obtained for pure PEBAX-4533 membranes in this study were 0.181, 0.36 and 0.246 Barrer at feed gas pressures of 3, 5 and 10 bar respectively. It is, thus, evident that PEBAX-1675 separation performance is superior to PEBAX-4533.

With incorporation of fillers (e.g. MWCNTs, nano-silica, H-mordenite and zeolites), different permeability values were obtained for PEBAX-1675. Overall, the separation performance was observed to generally enhance with the incorporation of fillers. In zeolite filled MMMs, the permeability coefficients increased with increasing both filler loading and the feed gas pressure [12,19-21]. Significant increase up to 464 Barrer was attained with zeolite imidazolate framework-8 (ZIF-8) at 2 Bar, followed by 758 Barrer at 8 bar [19,20]. Higher feed gas pressure enhanced the driving force for gas flux though the membrane which led to a higher extent of CO_2 adsorption. Additionally, both studies have found that after certain filler loading, a decrease in permeability values occurred. It was attributed to the effect of higher filler loadings that hindered the polymer chain mobility. This observation is in agreement with the permeation behavior of 3 wt% PEBAX-4533/HKUST-1 membrane, which showed a significant decrease in permeability. On the other hand, the permeability coefficient of pure PEBAX-1675 at 10 bar was observed to be as low as 1.1 Barrer [14]. Even with the addition of fillers (nano-silica and H-mordenite), the enhancement was insignificant.

Comparing the permeability coefficients of CuBTC MOF filled MMMs, it is observed that the membranes with CuBTC generally exhibit better separation performance than other fillers such as zeolites, CuBTC/zeolite/MIL-53 hybrid, etc. [7,22,23]. Comparing the polymers, matrimid CuBTC MMMs have been reported to exhibit higher permeability coefficients than PEBAX-4533/CuBTC MMMs. On the other hand, a study by Basu *et al.* [24] reported the permeability coefficient of 0.2 Barrer at 5 bar feed gas pressure for matrimid/CuBTC MMMs, which was lower than 0.36 Barrer observed in this study for PEBAX-4533/CuBTC MMMs at the same pressure.

4.4 Conclusions

In this study, defect-free MMMs were successfully developed using PEBAX-4533 with different weight fractions of HKUST-1 through solvent-evaporation method. The MMMs exhibited effective permeability for CO_2 gas at different pressure ranges. The permeability of N_2 as well as CH_4 in addition to CO_2/N_2 and CO_2/CH_4 mixed-gas systems is required to be investigated further to gain insights about the diffusivity and selectivity of the prepared MMMs.

References

1. Sridhar, S., Smitha, B., and Aminabhavi, T. M. (2007) Separation of carbon dioxide from natural gas mixtures through polymeric membranes - A review. *Separation and Purification Reviews*, **36**(2), 113-174.
2. Sheth, J. P., Xu, J., and Wilkes, G. L. (2003) Sold state structure property behavior of semicrystalline poly(ether-block-amide) PEBAX (R) thermoplastic elastomers. *Polymer*, **44**(3), 743-756.
3. Murali, R. S., Sridhar, S., Sankarshana, T., and Ravikumar, Y. V. L. (2010) Gas permeation behavior of PEBAX-1657 nanocomposite membrane incorporated with multiwalled carbon nanotubes. *Industrial & Engineering Chemistry Research*, **49**(194), 6530-6538.
4. Ovid'ko, I. A. (2013) Enhanced mechanical properties of polymer matrix nanocomposites reinforced by graphene inclusions: A review. *Review on Advanced Materials Science*, **34**, 19-25.
5. Venna, S. R., and Carreon, M. A. (2015) Metal organic framework membranes for carbon dioxide separation. *Chemical Engineering Science*, **124**, 3-19.
6. Li, H., Haas-Santo, K., Schygulla, U., and Dittmeyer, R. (2015) Inorganic microporous membranes for H_2 and CO_2 separation - Review of experimental and modeling progress. *Chemical Engineering Science*, **127**, 401-417.
7. Kanehashi, S., Chen, G. Q., Scholes, C. A., Ozcelik, B., Hua, C., Ciddor, L., Shouthon, P. D., D'Alessandro, D. M., and Kentish, S. E. (2015) Enhancing gas permeability in mixed matric membranes through tuning the nanoparticle properties. *Journal of Membrane Science*, **482**, 49-55.
8. Lim, S. Y., Choi, J., Kim, H.-Y., Kim, Y., Kim, S.-J., Kang, Y. S., and Won, J. (2014) New CO_2 separation membranes containing gas-selective Cu-MOFs. *Journal of Membrane Science*, **467**, 67-72.
9. Leal, O., Bolivar, C., Ovalles, C., Urbina, A., Revette, J., and Jose, J. (1996) *Carbon Dioxide Removal from Natural Gas using Amine Surface Bonded Adsorbents*. Online: https://web.anl.gov/PCS/acsfuel/preprint%20archive/Files/Merge/Vol-41 4-0003.pdf [accessed 15th February 2019].
10. *Synthetic Membranes*, Wikipedia (2017). Online: http://en.wikipedia.org/wiki/Synthetic_membrane [accessed 18th April 2017].
11. Shahid, S., and Nijmeijer, K. (2014) Performance and plasticization behavior of polymer-MOF membranes for gas separation at eleva-

ted pressures. *Journal of Membrane Science*, **470**, 166-177.

12. Murali, R. S., Ismail, A. F., Rahman, M. A., and Sridhar, S. (2014) Mixed matrix membranes of Pebax-1657loaded with 4A zeolite for gaseous separations. *Separation and Purification Technology*, **129**, 1-8.

13. Gue, X., Huang, H., Ban, Y., Yang, Q., Xiao, Y., Li, Y., Yang, W., and Zhong, C. (2015) Mixed matric membranes incorporated with amine-functionalized titanium-based metal-organic framework for CO_2/CH_4 separation. *Journal of Membrane Science*, **478**, 130-139.

14. Murali, R. S., Kumar, K. P., Ismail, A. F., and Sridhar, S. (2014) Nano-silica and H-mordenite incorporated poly(ether-block-amide)-1657 membranes for gaseous separations. *Microporous and Mesoporous Materials*, **197**, 291-298.

15. Lia, L., and Yang, G. (2009) Variable-temperature FTIR studies on thermal stability of hydrogen bonding in nylon6/mesoporous silica nanocomposite. *Polymer International*, **58**(5), 503-510.

16. Zhang, L., Ruesch, M., Zhang, X., Bai, Z., and Liu, L. (2015) Tuning thermal conductivity of crystalline polymer nanofibers by inter-chain hydrogen bonding. *RSC Advances*, **5**, 87981-87986.

17. Sun, B., Kayal, S., and Chakraborty, A. (2014) Study of HKUST (copper benzene-1,3,5-tricarboxylate, Cu-BTC MOF)-1 metal organic frameworks for CH_4 adsorption: An experimental Investigation with GCMC (grand canonical Monte-Carlo) simulation. *Energy*, **76**, 419-427.

18. Sorribas, S., Kudashiva, A., Scholes, C. A., Almedro, E., Zornoza, B., de la Iglesia, O., Tellez, C., and Coronas, J. (2015) Pervaporation and membrane reactor performance of polyimide based mixed matrix membranes containing MOF HKUST-1. *Chemical Engineering Science*, **124**, 37-44.

19. Jomekian, A., Behbahani, R. M., Mohammadi, T., and Kargari, A. (2016) CO_2/CH_4 separation by high performance co-casted ZIF-8/Pebax 1657/PES mixed matrix membrane. *Journal of Natural Gas Science and Engineering*, **31**, 562-574.

20. Xu, L., Xiang, L., Wang, C., Yu, J., Zhang, L., and Pan, Y. (2017) Enhanced permeation performance of polyether-polyamide block co-polymer membranes through incorporating ZIF-8 nanocrystals. *Chinese Journal of Chemical Engineering*, **25**(7), 882-891.

21. Mittal, G., Dhand, V., Rhee, K. Y., Park, S.-J., and Lee, W. R. (2015) A review on carbon nanotubes and graphene as fillers in reinforced polymer nanocomposites. *Journal of Industrial and Engineering Chemistry*, **21**, 11-25.

22. Zornoza, B., Tellez, C., Coronas, J., Gascon, J., and Kapteijn, F. (2013) Metal organic framework based mixed matrix membranes: An increasingly important field of research with a large application potential. *Microporous and Mesoporous Materials*, **166**, 67-78.

23. Li, W., Zhang, Y., Li, Q., and Zhang, G. (2015) Metal-organic framework composite membranes; synthesis and separation applications. *Chemical Engineering Science*, **135**, 232-257.

24. Basu, S., Cano-Odena, A., and Vankelecom, I. F. J. (2011) MOF-containing mixed-matrix membranes for CO_2/CH_4 and CO_2/N_2 binary gas mixture separations. *Separation and Purification Technology*, **81**(1), 31-40.

5

Micro-solid Phase Extraction using MOFs

Providencia González-Hernández,[1] Adrián Gutiérrez-Serpa,[1] Priscilla Rocío-Bautista,[1] Jorge Pasán,[2] Juan H. Ayala[1] and Verónica Pino[1,*]

[1]*Departamento de Química, Unidad Departamental de Química Analítica, Universidad de La Laguna, La Laguna (Tenerife), Spain*
[2]*Departamento de Física, Laboratorio de Rayos X y Materiales Moleculares, Universidad de La Laguna, La Laguna (Tenerife), Spain*

Corresponding author: veropino@ull.edu.es

5.1 Introduction

In the nineties, the research groups of Robson [1], Kitagawa [2], Yaghi [3] and Férey [4] developed porous crystalline coordination polymers, which were termed as metal organic frameworks (MOFs) by Yaghi and Li in 1995 [3]. MOFs present highly ordered three-dimensional structures. Their topology results from the combination of inorganic secondary building unit (SBU) and metal or metal ions cluster, coordinated with an organic linker (or linkers) through a specific connectivity [5]. Proper design of structures is possible, while ensuring the maintenance of a certain topology despite variations in the nature of the organic linker(s), if the coordination of the SBU is kept constant (isoreticular synthesis). From a theoretical perspective, there is almost an infinite number of possible combinations to form MOFs by varying the SBUs. More than 20,000 different MOFs have been registered in the Cambridge Structural Database (CSD); and many of them exhibit potential for a wide range of advanced applications [6,7].

Despite the significant interest in these highly stable materials, these crystals have not much to offer in terms of reactivity [5]. Indeed, the scientific attention received by MOFs, compared to more conventional porous materials (organic polymers or inorganic zeolites), is mainly associated to their structural characteristics and unique physico-chemical properties. Few of these include tuneability (possibility

Metal Organic Frameworks, edited by Vikas Mittal

of post-modifications and functionalization), high porosity (pore volumes up to 100 Å), stability (i.e. thermal resistance from ~200 to ~500 °C), ultra-low crystal densities (even down to 0.13 g·cm^{-3}), and impressive surface area (from 1000 to 10,000 m^2·g^{-1}) [6-7].

Many different routes have been described to prepare MOFs, all following crystallization approach (solvothermal, microwave-assisted, electrochemical, mechanochemical, sonochemical, layer-by-layer, etc.) using aqueous or non-aqueous media [7,8]. The MOF structure is highly dependent on its preparation method as small variations in the preparation method exert significant influence on the structure and properties of the obtained MOFs. In this sense, it is possible to control the resulting porosity and other features by modifying parameters during synthesis [6], such as the amount of MOF precursors (metal(s) and/or linker(s)), solvent nature and amount, temperature, time, type and amount of modifiers, size of the reactor, etc.

These outstanding materials have a plethora of applications. At the beginning, the main applications of MOFs were focused on gas storage [9,10], taking advantage of their exceptional porosity. Progressively, MOFs have been explored widely in biomedical [11,12], catalysis [13,14] and sensing [15,16] applications. MOFs incorporating biomolecules and/or biocompatible metals have also shown to be excellent materials in imaging and drug delivery [11,17]. Furthermore, luminescent MOFs have been proposed as potential materials in the electrochemical field [15,18].

Applications of MOFs in the analytical chemistry field have been more recent, and, in fact, their use in analytical separations has only been developed in the last decade. The use of MOFs in analytical chemistry has also been the topic of recent review articles in the field of (i) analytical sample preparation (with MOFs being used as sorbent materials) [19-21] and (ii) chromatography (with MOFs being used as novel stationary phases) [22,23]. Around 200 publications in the journals listed in the Journal Citation Reports (JCR) focus on the use of MOFs in sample preparation and/or chromatography [24]. Overall, the interest in MOFs is linked to the necessity of overcoming problems of common sorbents in various applications.

5.2 Analytical Sample Preparation: Microextraction Strategies and Novel Materials

Direct sample analysis is not always possible despite the use of modern mass spectrometers (MS), particularly when complex solid and/or fatty samples are handled (thus needing clean-up steps and removal of interferences) or when the analytes to be determined are present in very low concentrations (thus needing pre-concentration strategies). Sample preparation is, therefore, a crucial stage in analytical methodologies.

Most conventional analytical sample preparation methods include solid-phase extraction (SPE) and liquid-liquid extraction (LLE). Unquestionably, there are a number of advantages of the conventional SPE *versus* the classical LLE, including lower organic solvent consumption, elimination of the risk of the emulsion formation, simplicity and feasibility of automation. Owing to these advantages, SPE is the most frequently used extraction procedure in sample preparation [19,25].

Trends in sample preparation have shifted to the development of methods that permit the minimization of the sample amounts or at least the minimization of the need of the organic solvents during the sample preparation, along with the reduction in the chemical wastes, cost and time. It is possible to compile with the principles of the green analytical chemistry (GAC), if several of the abovementioned requirements are fulfilled. In this sense, SPE has been the subject of a number of miniaturization approaches.

Advances in sorbent-based microextraction techniques include the main modes: solid-phase microextraction (SPME), which requires a coated fiber [26,27] and miniaturized solid-phase extraction (µSPE), which does not require a coted fiber. They also include a number of sub-modes, such as stir-bar sorptive extraction (SBSE), stir-cake sorptive microextraction (SCSE), hollow fibers (HF) to protect SPME fiber coatings, miniaturized devices containing sorbents, among others. These techniques have been undoubtedly expanded by the incorporation of novel materials, thus, demonstrating improved performance over conventional media [25,28,29].

SPME, SBSE and SCSE normally require larger extraction times than µSPE approaches. It is true that the requirements for the SPME, SBSE or SCSE sorbents are as restrictive as any material used in µSPE, including high mechanical, chemical and thermal stability as well as good selectivity and adsorptive/absorption capacity, along with rapid kinetic for the interactions with the target analytes, if possible [20,29]. However, ease of preparation of a stable coating strongly bonded to the inert support of the fiber is needed in SPME (that somehow limits its expansion). Indeed, the number of commercially available SPME coatings around six, only commercialized by Merck [30] and Restek [31]. In fact, considering the applications of MOFs in sorbent-based microextraction approaches, there is a higher number of applications in µSPE [19] than in SPME [20].

The versatility and unique characteristics of MOFs as compared to other sorbent materials (including novel materials such as molecularly-imprinted polymers [32], metallic nanoparticles [33] and carbonaceous materials such as fullerenes, carbon nanotubes or graphene [29]) explain their potential as (almost ideal) sorbents for solid-based microextraction methods in analytical sample preparation. To date, more than 100 studies have been published on applications of MOFs in µSPE, whereas roughly 40 publications report MOF-based SPME coatings and their applications.

5.3 MOFs in µSPE Techniques

The µSPE analytical sample preparation technique can be classified into two main modalities: the static-conventional mode and the dispersive mode, referred to as conventional µSPE and D-µSPE in this chapter, respectively. In the conventional static approach of µSPE, the solid sorbent is confined in a device and the sample flows through it to achieve the partitioning of the analytes present in the sample. The analytes retained by the sorbent are subsequently subjected to elution or desorption to accomplish their analytical determination. In the D-µSPE mode, the solid sorbent is directly added to the sample while being subjected to a strong stirring to ensure uniform transfer of the analytes to the sorbent. After separation of the sorbent containing trapped analytes, further steps of elution or desorption are also carried out.

In addition, there is a version of the D-µSPE mode that uses magnetic materials as sorbents, implying an improved microextraction procedure facilitated by magnets (separation of the magnetic material is achieved by the application of an external magnetic field), referred to as M-D-µSPE in this chapter. MOFs as sorbent materials have been successfully used in all of the mentioned versions of µSPE for the extraction of different analytes from complex sample matrices. Table 5.1 show the reported applications of MOFs in µSPE [34-48], D-µSPE [49-91] and M-D-µSPE [91-148]. In these applications, it can be observed that the most frequently employed MOFs as sorbents are MIL-101(Cr) and HKUST-1. Apart from these two MOFs, only MOF-5(Zn), MIL-53(Al) and UiO-66(Zr) have been applied in all of the modalities of µSPE. To show a more comprehensive evaluation of the

Table 5.1 Overview of analytical applications of MOFs (and several of their functionalized derivatives) as sorbents in the different micro-solid phase extraction modes.

MOF	μSPE	D-μSPE	M-D-μSPE	Generic sample (types)	Analytes (number)
CD-MOF	–	[79]	–	food (chicken, pork, liver and fish)	SAs[a] (5)
CYCU-3	[41]	–	–	water	drugs (9)
DUT-4	[41]	–	–	water	drugs (9)
DUT-5	[41]	–	–	water	drugs (9)
Dy-MOF	–	[68]	–	water	metal ions (2)
HKUST-1	[36]	[54,59,63,83-85,82]	[93,95-97,102-104,106,110,112,113,115,131]	biological (exhaled breath, urine), personal care products (body cream), food (baby food, basil, beetroot, coriander, fenugreek, fish, leek, milk, peanuts, parsley, tea), sediments, soil, water (tap, river and waste)	aldehydes (6), drugs (6), disinfectant (1), dye (1), flavonoids (1), herbicides (7), hormones (2), metal ions (7), PAHs[b] (9), parabens (7)
IRMOF-3	–	–	[100,148]	food (celery, pakchoi and spinach), sediments, water (lake, river, irrigation and reservoir)	herbicides (4), metal ions (1)
JUC-62	–	[69]	–	food (mushroom and tea)	metal ion (1)
JUC-48	–	–	[119]	food (chicken, pork and shrimp)	SAs[a] (5)
MIL-100(Cr)	–	[81]	[101]	biological (urine), food (fish), water (aquaculture, groundwater, river, tap)	drugs (8), dye (1)
MIL-100(Fe)	[40]	[52,67,88,117, 54,55,81,90]	[134,140, 141,143]	biological (urine), food (apple juice, carrot, grape juice, lettuce, milk, and potato), personal care	disinfectant (1), drugs (10), EDCs[c] (6),insecticides (1), PAEs[d] (6), PAHs[b] (13),

				product (, baby toilet water, toothpaste), water (feed water, mineral, lake, pond, river, snow, tap, wastewater, well)	PCBs[e] (7), SAs[a] (12)
MIL-101(Cr)					
MIL-101(Cr)	[34, 39,40, 42,48]	[51-56,58-60,64-67, 75,89,90]	[92,116, 123,128-130,132]	biological (urine), food (apple, chicken, garlic, milk, onion, orange, peanuts, rice, soybeans, soy oil, tea, vegetable oil), personal care products (cosmetics), toner (ink), and water (feed water, fishpond, mineral, lake, reservoir, river, sea, tap, and wastewater)	benzophenones (3), BP-UV filters, drugs (23), EDCs[c] (5), flavonoids (3), hormones (12), insecticides (4), herbicides (17), metal ions (2), PAEs[d] (6), PAHs[b] (10), PCBs[e] (7), pesticides (11), SAs[a] (13)
MIL-101(Cr)-NH₂	–	–	[118]	drugs (irbesartan active pharmaceutical)	azides (2)
MIL-101(Fe)					
MIL-101(Fe)	–	–	[99,105, 108,109, 122,127, 148]	biological (basil, beetroot, coriander, fenugreek, garden cress, hair, leek, parsley, radish, and urine), food (canned tuna, celery, fish, pakchoi, shrimp, spinach, and tea), and water (irrigation, mineral, lake, rain, reservoir, river, sea, wastewater)	fungicides (4), herbicides (4), metal ions (5), pesticides (6)
MIL-101(Fe)-NH₂	–	–	[127]	water (lake, river)	fungicides (4)
MIL-125(Ti)	–	[91]	–	water (mineral, snow, among others)	NPAHs[f] (16)

MIL-53(Al)

MIL-53(Al)	[38]	[54,55,63,81,82]	[107]	biological (urine), personal care products (body cream), water (aquaculture, feed water, lake, river, tap, and wastewater)	disinfectant (1), drugs (9), EDCs[c] (6), PAHs[b] (1), PCBs[e] (7)
MIL-53(Al)-NH$_2$	–	[55]	–	biological (urine)	EDCs[c] (4)

MIL-53(Fe)

MIL-53(Fe)	–	–	[127]	water (lake, and river)	fungicides (4)
MIL-53(Fe)-NH$_2$	–	–	[127]	water (lake, and river)	fungicides (4)
MIL-68(Al)	[45]	–	–	food (mollusks)	metal ions (2)
MOF(Ag)$_n$	–	[86]	–	soil (orchard soil)	polybrominated diphenyl ethers (7)
MOF(B)	–	–	[147]	biological (human serum digest)	glycopeptides (209)
MOF-177(Zn)	–	–	[98]	water (lake, river, and wastewater)	phenols (6)
MOF-5(Zn)	[35, 37]	[63,73,82]	[94,121, 125,126]	biological (urine), food (fish, an seaweed), personal care products (body cream), water (reservoir, river, and tap)	disinfectant (1), drugs (5), gibberellic acid (4), hormone (2), NPAHs[f] (2), PAHs[b] (11), PCBs[e] (7), pesticides (4)
MOF-545(Zr)	[43]	[74]	–	food (beans, chickpea, cherry juice, fish, lentil, and wheat), water (mineral, wastewater, and well)	metal ions (2)
TMU-4	–	[61]	–	water (mineral, river, and tap)	metal ions (5)

TMU-5	–	[61]	–	water (mineral, river, and tap)	metal ions (5)
TMU-6	–	[61,70]	–	water (mineral, river, and tap)	metal ions (5), and PAEs[d] (3)
TMU-8	–	–	[124]	food (apples, carrots, cucumbers, onions, tea, tomatoes, and watermelon), and water (river, and tap)	metal ions (7)
TMU-9	–	–	[124]	food (apples, carrots, cucumbers, onions, tea, tomatoes, and watermelon), and water (river, and tap)	metal ions (7)
UiO-64(Zr)	–	[82]	–	water (tap, and wastewater)	hormone (2), drugs (2), disinfectant (1), and PAHs[b] (1)
UiO-66(Zr)					
UiO-66(Zr)	[46, 47]	[78,55,81]	[135,136]	biological (human serum, and urine), food (milk, mung, shellfish and watermelon), and water (aquiculture, lake, river, tap, and wells)	domoic acid (1), EDCs[c] (6), hormones (4), insecticides (6), phenols (5), phosphopeptides (6)
UiO-66(Zr)-NH₂	–	[71]	[138,139]	biological (human serum), soil, water (mineral, and river)	PCBs[e] (7), phenols (6), SAs[a] (2)
UiO-66(Zr)-NH-CH₃	[44]	–	–	biological, (human serum), and food (milk)	phosphopeptides (6)
UiO-66(Zr)-OH	–	[80]	–	water (reservoir)	metal ions (1)
UiO-66(Zr)-SO₃H	–	–	[142]	biological (human serum)	glycopeptides (177)
UiO-67(Zr)	–	[72]	–	food (grapefruit, and pear)	plant growth regulators (8)

ZIF-11(Zn)	–	[57]	–	water (river, spring, and tap)	PAHs[b] (7)
ZIF-67	–	–	[120]	food (lily)	aromatic amino acids (3)
ZIF-7(Zn)	–	[57]	[111]	air, water (rainwater, river, spring, and tap)	PAHs[b] (8)
ZIF-8(Zn)	–	[49,50,54, 77,87]	[137,144, 145]	biological (urine), plants (*Caragana Jubata*), water (canal, feed water, fishpond, pond, river, spring, tap, and wastewater)	drugs (5), flavonoids (9), fungicides (5), PAHs[b] (6), peptides, phenols (2)
Zn-BTC	–	[76]	[133]	water (tap, river, and wastewater)	drugs (1), pesticides (6)
ZnCB	–	–	[146]	biological	drugs (9)

[a]sulphonamides
[b]polycyclic aromatic hydrocarbons
[c]endocrine disrupting chemicals
[d]phthalic acid esters
[e]polychlorinated biphenyls
[f]nitro-polycyclic aromatic hydrocarbons

applications of MOFs as sorbents and to provide an overview on the current trends, Figure 5.1 shows the number of studies published since the first use in µSPE (Figure 5.1(A)) and the percentage of the overall applications devoted to the three modalities (Figure 5.1(B)).

5.3.1 MOFs as Sorbents in the Conventional µSPE Mode

The use of MOFs (bare MOFs or hybrid materials incorporating MOFs) in the conventional static µSPE approach (Figure 5.2(A)) has been accomplished by confining them in SPE cartridges [35,40,42] or in other type of devices [36,37,43,45,47,48], including the incorporation of MOFs in monolithic microcolumns [38,41,44,46]. Furthermore, MOFs have been employed in online µSPE [34,39]. Figure 5.2(B) shows main MOF-based devices developed to date for static conventional µSPE. Only two neat MOFs have been packed in SPE cartridges so far: MOF-5(Zn) [35] and MIL-101(Cr) [40]. Recently, Li *et al.* [42] reported the use of a composite formed by MIL-101(Cr) and a molecularly-imprinted polymer in a µSPE cartridge [42].

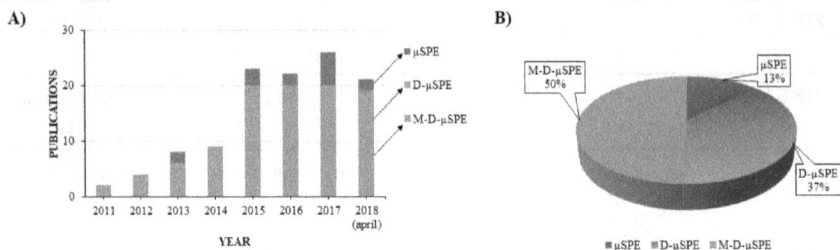

Figure 5.1 A) Publications reporting the use of MOFs as sorbent materials in µSPE, as a function of the year of publication and µSPE mode and B) distribution of the publications (as overall percentage) devoted to the different µSPE modes.

MOFs in online µSPE have been used in coupling with high performance liquid chromatography (HPLC) using MIL-101(Cr) as sorbent [34,39]. In the first study, MOF was packed in a drilled tube of polyetheretherketone (12×7.2 mm) [34] and in the second work, the sample loop was replaced by a stainless steel column (25×4.6 mm) filled with the MOF material [39].

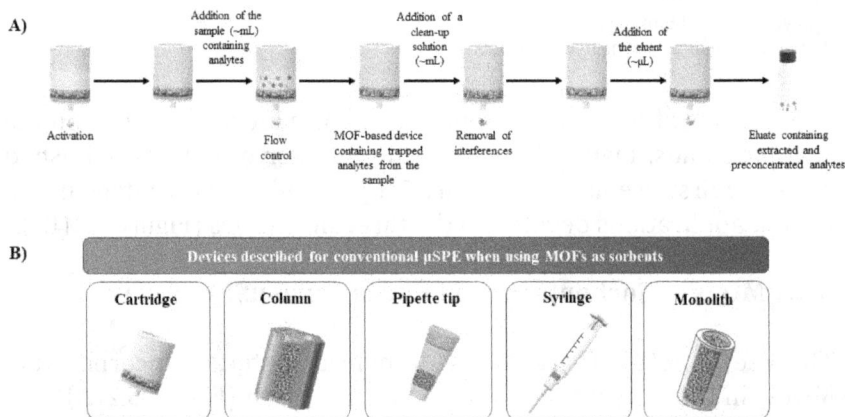

Figure 5.2 A) General scheme of the static micro solid-phase extraction procedure and B) main miniaturized devices described to perform static and conventional miniaturized extraction with MOFs as sorbents.

Syringe-based devices have also been fabricated containing MOFs. Asiabi *et al.* [37] prepared a disk containing nanofibers of the composite material formed by MOF-5 and polyacrylonitrile (MOF-5/PAN). The disk was incorporated to a syringe with 1 mL of capacity, fixed with frits [37]. The authors also fabricated a lab-scale filter employing a MOF-based composite as sorbent [45]. The authors fixed a nanofiber material composed of MIL-68(Al)/chitosan into the filter, requiring only 6 mg of MIL-68(Al). An easier packing procedure was developed by Zhang *et al.* [48] by directly packing 2 mg of MIL101(Cr) in the nozzle of a disposable syringe [48]. Other devices include the stainless steel mesh (13 mm×0.14 mm, with an average pore of 75 μm) coated with HKUST-1 (3 mg) by Wang *et al.* [36]. Pipette tips containing MOFs have also been also reported by using the neat MOF-545(Zr) [43] or the UiO-66/polyacrylonitrile composite [47].

With regards to monolithic columns, the MOF-based composite is normally synthetized by polymerization of two monomers in the presence of the selected MOF. For instance, Lyu *et al.* [38] fabricated a monolithic column based on MIL-53(Al)@poly(BMA-EDMA) in a silica capillary (2 cm long and 0.53 mm internal diameter (i.d.)). Using the same polymer but a different MOF, Shih *et al.* [41] demonstrated that the internal diameter of the capillary had an influence on the thermal and chemical stability of the monolithic DUT-5@poly(BMA-EDMA) composite. In another study, Li and Bie [46] fabricated the UiO-66@poly(MAA-PEGDA) monolithic (0.15 mm i.d.) structure using the same approach. The authors also fabricated a MOF-based monolith containing the amino derivative of UiO-66 and PEGDA as monomers. Firstly, UiO-66-NH$_2$ was modified with methacrylic anhydride, and the monolith was subsequently prepared through free radical copolymerization reaction. Poly(UiO-66-NH-Met-PEGDA) monolithic columns (0.15 mm i.d.) were successfully used in μSPE [44].

5.3.2 MOFs as Sorbents in the Dispersive Versions of μSPE

Main advantages of D-μSPE over conventional static μSPE include lower extraction times, higher extraction efficiency linked to the strong analyte-sorbent interaction and simplicity of the entire procedure [149]. Depending on the nature of the material used in D-μSPE, it is possible to distinguish between D-μSPE (Figure 5.3(A)) and M-D-μSPE (Figure 5.4(A)), the latter avoiding centrifugation and/or filtration steps as the magnetic material is easily separated from the sample matrix and remaining non-extracted components.

MOFs in D-μSPE

More than 25 different MOFs have been employed as sorbents in D-

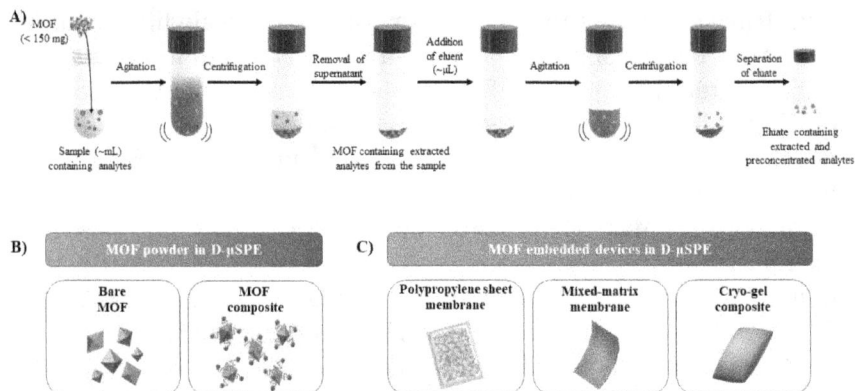

Figure 5.3 A) General scheme of the dispersive miniaturized solid-phase extraction procedure, B) MOF powder used in D-μSPE and C) MOF embedded devices employed in D-μSPE.

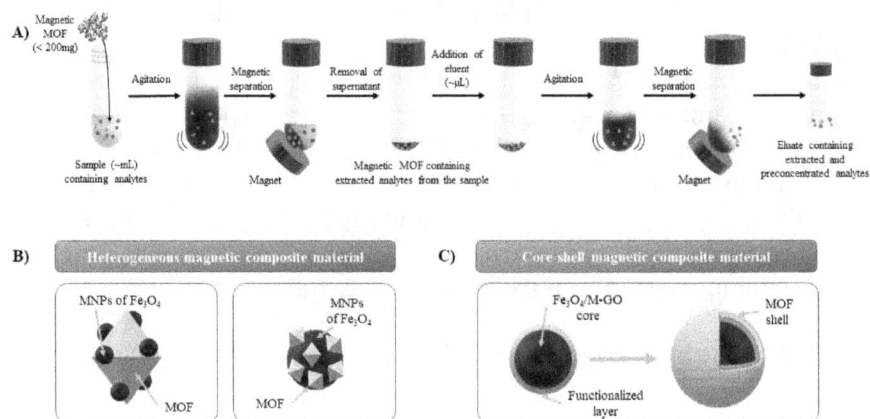

Figure 5.4 A) General scheme of the magnetic dispersive miniaturized solid-phase extraction procedure, B) heterogeneous magnetic MOF composites used in M-D-μSPE and C) core-shell magnetic MOF composites employed in M-D-μSPE.

μSPE (Table 5.1). In the reported applications, an activated MOF is added to the sample (normally an aqueous sample, with volume between 1 and 20 mL), with amount of MOFs ranging from 0.5 to 150 mg. The most common stirring modes include manual agitation [58,59,68,69,76,80,83,84,87], magnetic stirring [49-53,61,70,85,86], vortex-assisted agitation [54,55,63,65-67,71,72,74,75,78,81,82,88-

90] and ultrasound [56,60,62,64,73,91]. The vigorous stirring favors strong interactions between the MOF and target analytes, which, in turn, ensure higher extraction efficiencies in D-μSPE. Moreover, extraction times are quite short (20 min on average). After separation of the MOFs containing the extracted analytes (by centrifugation, filtration or decantation), a low volume of eluent (in the μL range) is normally added to ensure removal of analytes from the MOF. Many studies have reported the use of ultrasound as the stirring mode to ensure the maximum elution efficiency [49-54,56,59,60,62,64-67,71-73,75,81,88-91]. In other cases, the same stirring mode used in the extraction step is also utilized during elution [55,63,69,74,78,80,82,84-86]. The time required to ensure an efficient elution is generally 15 min [56,75]. Depending on the compatibility of the elution solvent with the analytical technique, additional solvent-exchange steps are required in some cases [51,54,56,57,59,60,62-67,72,75,78,81,82,89-91].

Two kinds of MOFs-based materials have been employed in D-μSPE, MOF as powders (including neat crystals or MOF-based composites) and MOF embedded devices (Figure 5.3(B) and (C)). The main difference between the two approaches lies in the dispersion capability. When using MOF powders, each crystal acts as an independent unit with free movement, and the entire surface gets in contact with the sample. When utilizing MOFs embedded devices, crystals are retained in (or immobilized onto) another material. Thus, their dispersion is more limited and controlled, and the accessibility of the sample to the entire material requires longer times.

Use of MOF Powders in D-μSPE

Since the study of Ke *et al.* [83] describing the utilization of MOF (HKUST-1) powder as the solid dispersant media for the removal of mercury from water [83], MOFs have gained an increasing role as potential sorbents in D-μSPE. These MOFs powders include the neat crystals or composite materials based on MOFs, as shown in Figure 5.3(B).

The most common neat MOF used as powder in D-μSPE is MIL-101(Cr), formed by chromium (as metal ion) and terephthalic acid (as organic ligand) [56,58-60,64-67,75]. MIL-101(Cr) has become widely used due to its properties, such as accessible coordinative unsaturated sites which improve the interactions with the analytes presenting electronegative groups in their chemical structure [58]; large pore windows (1.2 and 1.47×1.6 nm) and mesoporous pores (2.9 and 3.4 nm), which allow the analytes to enter into the crystal structure [56,59,66,67]; Brunauer-Emmett-Teller (BET) surface area values around 3023 $m^2 \cdot g^{-1}$; good chemical stability [56,59,66,67], etc. It is also important to mention that this MOF is commercially available, thus, facilitating its testing for a wide variety of applications. However, MIL-101(Cr) is unstable in alkaline media, particularly at pH values higher than 9.6 values [56,65-67,75].

Other successful MOFs powders in D-μSPE include UiO-66(Zr) [78] and MIL-53(Al) [81,82]. MIL-53(Al) also presents coordinative unsaturated sites, thus, enhancing the possible interactions with target analytes [82]. It has a flexible structure (it is a breathing MOF, with adjustable structure to guest analytes) that probably compensates the relatively lower surface area (around 978 $m^2 \cdot g^{-1}$) [81]. In the case of UiO-66, it does not present unsaturated metal sites, the surface area is even lower (around 669 $m^2 \cdot g^{-1}$), and the pore volume is ~0.52 $cm^3 \cdot g^{-1}$. [78,81]. Nevertheless, this MOF has an extraordinary functionalization feasibility, allowing the design of the organic linkers with better interaction capabilities with the target compounds, such as UiO-66-OH [80] and UiO-66-NH$_2$ [71]. These MOFs have missing-linkers defects, leading to enhanced porosity. Among the UiO MOFs family, not only UiO-66(Zr) and derivatives, but also UiO-67(Zr) has been tested in D-uSPE applications with promising results [72].

Other MOF family worth mentioning is the TMU, particularly TMU-5 [61] and TMU-6 [70], which can be prepared by mechano-chemical synthesis (organic solvent free) and exhibit resistance to alkaline environments.

Composites based on MOFs have been also utilized as powders in D-μSPE [85,87-91]. Graphene oxide (GO) is an excellent candidate to be decorated with MOFs. When GO is combined with MOFs, the resulting material presents stability in water, along with increased sorption capability (if compared with the neat MOF), better dispersion into the sample, larger pore volume and slightly higher BET surface area [89,90]. The preparation of these composites is usually simple, with MOFs precursors added to the GO colloidal suspension before crystallization [85,89,90]. The resulting material implies that the GO structure is decorated with MOF crystals, but the MOF crystals do not suffer incorporation of GO in the framework [89,90].

Other materials useful for forming composites with MOFs and used as powders in D-μSPE include the macroporous resin D101 [87] and silica gel [86].

Other interesting materials used in MOF-based composites are the ionic liquids (ILs). The composite formed by [BMIm][Cys] and MIL-100(Fe) has been observed to be particular relevance [88]. In this case, the MOF is firstly prepared by a solvothermal method, and the crystals are subsequently modified with [BMIm][Cys]. The IL gets inside the pores or close to the pore openings, leading to a decrease in the overall surface area and micropore volume of the final composite, but providing better adsorption capabilities (owing to the well-known solvation abilities of ILs with target analytes).

Use of MOF Embedded Devices in D-μSPE

MOF embedded devices are formed by a main rigid (or semi-rigid) body containing MOFs. Main advantage of their use is the possibility of performing the extraction procedure without the need of any centrifugation, filtration and/or decantation steps to separate the MOF (containing trapped analytes) from the sample after extraction (or from the eluent after the elution step). The devices can be easily removed from the sample (and even from the elution solvent), thus, reducing the number of steps and sources of errors. These devices can be divided into three main groups based on their composition and configuration: devices prepared with polypropylene (PP) sheet membranes [49,50,52,53], devices prepared with MOF mixed-matrix membranes (MOF-MMM) [55] and devices prepared with a cryogel composite [54]. Figure 5.3(C) shows the MOF embedded devices mainly used in D-μSPE. Overall, most of the devices utilize PP sheet membranes.

PP sheet membranes can be used to form rectangular prism devices with dimensions around 0.5-1.0×0.5-1.5 cm. These devices are prepared by folding the membranes followed by heat-sealing of the two (out of three) open sides. Afterwards, the MOF powder is introduced through the open side of the device. Finally, the open side is also heat-sealed to obtain confined MOF powder surrounded by the PP sheet membrane. So far, the MOFs employed in these devices are MIL-101(Cr) [52,53] and ZIF-8(Zn) [49,50]. The amount of MOF powder required in these devices depends on the size of the final device, being around 4-5 mg for the smaller devices [49,50,53], while larger devices require around 10 mg [52].

Zang *et al.* [51] also proposed a variation of this strategy by confining the MIL-101(Cr) powder in the hollow fibers instead of using PP sheets.

Main applications of the MOF-MMM devices can be found in gas sorption applications, despite being an interesting option for D-μSPE [55]. The device is prepared by generating an ink mixture containing 150 mg of (dried) MOF dispersed in ethanol and polyvinylidene difluoride/dimethylformamide solution. Such MOF-ink hybrid is subsequently distributed homogenously onto a glass substrate, followed by heating to remove the solvents. This procedure ensures the formation of a thick membrane film. The color of the membrane depends on the nature of the MOF used in the ink media (brown membranes for MIL-101(Fe), green for MIL-101(Cr), yellow for UiO-66(Zr), etc.) [55].

MOF cryogel composites are the novel materials reported by Wang *et al.* [54] (specifically a MOF polyvinyl alcohol cryogel (MOF/PVA cryogel)) for use in the devices for D-μSPE. Cryogels are the materials with low surface areas and low adsorption capability. However, these also contain interconnected macropores that allow analytes an easy diffusion path through the material. The introduction of MOFs into the cryogels improves the material adsorption capability [54]. The preparation of MOF cryogel composites is similar to MOF-MMM. Briefly, a mixture of the PVA solution and MOF powder is placed on a glass sheet at low temperatures (-20 °C) for 12 h. Once the cryogel monolith is formed, it is thawed and washed several times before use.

MOFs in M-D-μSPE

Magnetic composites based on MOFs require the combination of MOF crystals with a magnetic material. The general scheme of the magnetic-facilitated microextraction procedures, M-D-μSPE, when using magnetic-composites based on MOFs, is shown in Figure 5.4(A). The amount of the magnetic material required in M-D-μSPE is slightly higher than D-μSPE, but lower than 200 mg in any case. The stirring is normally accomplished using ultrasound [92,94,98,106,107,110, 116,119,121,126,127,129,132,134,137,139,140,146], vigorous shaking [93,95-97,99,100,108,117,118,123,126,133,142,144,145], stirring [101-105,107,111-113,120,122,128,130,131,138] and vortex [109,115,135,148] for average time of 10 min, however, lower than 40 min in any case. The advantage of this procedure lies in the ease of the separation of the MOF material, thus, avoiding the use of additional centrifugation and/or filtration steps, as the simple action of an external magnet is sufficient. This simplicity justifies the large number of applications developed using this microextraction mode (Figure 5.1(B)). In most cases, separation step only requires few seconds, owing to the easy decantation of the supernatant. The same approach is accomplished after elution: the eluting solvent is separated from the magnetic MOF with the aid of the external magnet.

Maya *et al.* [101] proposed a variation in this general M-D-µSPE procedure: the magnetic MOF composite is placed on a stir bar, and this device is placed into a syringe. The device permits the automation of the M-D-µSPE procedure, thus, offering an attractive alternative for routine analysis.

Owing to their morphologic features, magnetic-based MOF composites can be classified into two groups: heterogeneous magnetic composites (Figure 5.4(B)) [92-132] and core-shell magnetic composites (Figure 5.4(C)) [133-148].

Use of Heterogeneous Magnetic MOF Composites in M-D-µSPE

Heterogeneous magnetic MOF composites are formed by the combination of MOF crystals with at least one magnetic material following a non-homogeneous distribution (Figure 5.4(B)). Further, other additional materials are commonly incorporated into the heterogeneous magnetic MOF composites such as GO [131,132] or multi-walled carbon nanotubes (MWCNTs) [120].

Magnetite (α-Fe$_3$O$_4$) is the most commonly used magnetic material, with different size variants, such as nano-particulates [93,94,99,101,104,119], micro-particulates [92,95,98,106,116,118] and particulates [100,127]).

The incorporation of magnetite to the MOF materials to form the heterogeneous composites is achieved by an *in-situ* physical blending (by adding the magnetite to a solution containing the dispersed MOF). The magnetite interacts with the metal ions of the MOF's SBUs, thus, resulting in immobilization [92,98,106,110,111,116-118,132]. This option is quite attractive due its simplicity. Nevertheless, the physical blending interactions between the MOF and magnetite are not always as strong as desired, and the weakly attached MOF-magnetite structure, in general, justifies the low stability of the materials prepared using this procedure. Other studies have described the preparation of the magnetic composites by chemical bonding [94-97,99-101,112-115,119,120,122]. To obtain such composites, one of the two materials is synthetized previously and is dispersed in the medium in which the other is synthetized. Therefore, it is possible to prepare the MOFs first, followed by dispersion into a solution containing magnetic nanoparticle precursors [114,121,128,129]. It is also possible to prepare the magnetic nanoparticles first, followed by their dispersion into a solution containing the MOF precursors [92-100,102-109,112,113,115,119,120,122,124-127,130-132]. The second option has been commonly used in the literature, as it allows further functionalization of the magnetic nanoparticles prior to their coupling with MOFs crystals.

Most of the studies reported so far dispersed neat Fe_3O_4 nanoparticles [95,99,100,104,107,124,127] or mixtures of Fe_3O_4 with MWCNTs [120]. Nevertheless, it is important to take into account that the magnetic nanoparticles can form agglomerates and suffer degradation due to the chemical instability in the acidic media, which is required for the synthesis of several MOFs such as MIL-101(Cr). Thus, the preparation of coated or functionalized magnetic nanoparticles might avoid these problems by enhancing the stability of the magnetic material. Furthermore, the functionalization of Fe_3O_4 nanoparticles might also help to ensure stronger chemical bonding between the MOF and magnetite. Examples of common coatings for magnetite include silanization with tetraethyl orthosilicate (TEOS) [92,98], 3-chloropropyltriethoxysilane (3CPS) or (3-glycidyloxypropyl)-trimethoxysilane [112]. The layer can be further functionalized with 3-aminopropyl triethoxysilane (APTES) [94,131] or with chelating agents such as dithiocarbamate [115,130], dithizone [96], 2,5-dimercapto-1,3,4-thiadiazole [102,105] and 4-(thiazolylazo) resorcinol [113]. The chelating agents retain metal ions (the MOF precursor) on the surface of the magnetic nanoparticles, thus, forming complexes. Hence, the metal retention favors the binding of the magnetite with the MOF crystals.

In several cases, a prior TEOS layer is required for the functionalization. Thus, silanization and amino-functionalization can take place at the same time by coating the nanoparticles with *N*-[3(triethoxysilyl)propyl]isonicotinamide [108,122] or *N*-[3(triethoxysilyl)propyl]isonicotinamide [93,97]. Other option is to perform a direct mercapto-functionalization on the surface of the magnetic particles with mercaptoacetic acid, thus, complexing the metal precursor of the MOF material [109,119].

To minimize the preparation time of heterogeneous magnetic MOF composites, Bahrani *et al.* [121] proposed a combination of the *in-situ* and chemical bonding [121] strategies. In this approach, the MOF crystals and magnetite are mixed (as in the *in-situ* physical blending), however, an encapsulation step with ethylene glycol is performed to ensure magnetite's presence over the MOF surface.

It is important to highlight that irrespective of the way magnetic nanoparticles are attached to the MOF, the BET surface area of the resulting magnetic material is lower than the neat MOF. Therefore, though the magnetic composites provide operational simplicity for the microextraction process, however, higher efficiencies are obtained in D-μSPE.

Among the heterogeneous magnetic MOF composites mostly used in M-D-μSPE, it is important to highlight the HKUST-1 composites (HKUST-1@Fe_3O_4), as shown in Table 5.1. These were first proposed by Bagheri *et al.* [93] as chemically bonded composites [93] and by Rocío-Bautista *et al.* [110] using an *in-situ* approach [110]. In several applications, the HKUST-1@Fe_3O_4 composites have been functionalized with dithizone [95]. However, in general, the HKUST-1 composites have been used without further coating or functionalization [93,97,102-104,106,110,112,113,115]. MIL-101(Cr) [92,116,123,128-130,132], and MIL-101(Fe) [99,105,108,109,122] have been also widely used in M-D-μSPE applications. BET surface area of MIL-101(Fe) is lower (2328 $m^2{\cdot}g^{-1}$) than MIL-101(Cr), however, its pore volume (1.31 $cm^3{\cdot}g^{-1}$) is higher than MIL 101(Cr) (0.83 $cm^3{\cdot}g^{-1}$) [92,105,122]. Other MOFs employed for these applications include zinc-based MOFs such as MOF-5(Zn) [94,121,125,126], IRMOF-3 [100] and MOF-177 [98] as well as zeolitic imidazolate frameworks such as ZIF-7 [111] and ZIF-8(Zn) [137,144,145].

Use of Magnetic Core-Shell MOFs in M-D-μSPE

Core-shell materials are the materials composed of at least two easily distinguished parts, an internal material forming the center (core) and an external material coating the core (shell). If the core is not completely coated by the shell, the structure is not considered as core-shell material. Furthermore, the core usually presents paramagnetic properties. Figure 5.4(C) shows the schematics of the representative core-shell materials. Core-shell magnetic MOF composites have been commonly prepared with magnetic Fe_3O_4 microparticles obtained by co-precipitation [133], Fe_3O_4 microparticles obtained by the solvothermal method [134,135,137-148] or magnetic GO particles [136] as the cores.

Nooreini *et al.* [133] were the first to describe a procedure to obtain core-shell magnetic MOF composites. For this purpose, Fe_3O_4

magnetic nanoparticles were coated with TEOS to form a porous silica layer. Subsequently, $Fe_3O_4@SiO_2$ microparticles were functionalized with APTES, followed by the formation of a carboxyl-activated surface through the introduction of an adequate organic linker (such as furan-2-carboxaldehyde). Finally, functionalized particles were dispersed into a solution containing MOF precursors, and the MOF synthesis resulted in the coating of the particles with Zn-BTC [133].

Several studies have followed the approach to obtain core-shell structures by changing the organic ligand used in the carboxylation step. Jia *et al.* [139] proposed the use of succinic anhydride to obtain UiO-66 core-shell magnetic microparticles [139]. Glutaric anhydride has also been successfully used with UiO-66 [135] and ZIF-8 [137,144]. Recently, Ma *et al.* [146] substituted the anhydrides with GO using the Hummer's method [146].

Although the synthetic protocol reported by Nooreini *et al.* [133] has led to adequate results, it is tedious and time-consuming. Also, it does not ensure that a core-shell structure is obtained in all cases. Chen *et al.* [134] developed an alternative method by functionalizing Fe_3O_4 microparticles with mercaptoacetic acid, followed by the MOF synthesis. This strategy has been used for the synthesis of MIL-100(Fe)@Fe_3O_4 core-shell composites [134,140]. Yang *et al.* [141] also proposed the use of thioglycolic acid instead of mercaptoacetic acid. This functionalization has been applied to obtain MIL-100(Fe)@Fe_3O_4 core-shell composites [141,143].

Another alternative is to coat the magnetic microparticles with a polymer. Such a polymer coating has the same function as the silica layer, and it does not require further functionalization. A mixture of polyethylenimine/polyvinyl pyrrolidone [147] has been commonly used for this purpose. Polydopamine coatings have also been widely used using a self-polymerization approach [136,138,142,145,148].

Overall, the magnetic core-shell MOF composites commonly used in M-D-µSPE have been based on UiO-66 and its derivatives [135,136,138,139,142], followed by MIL-100(Fe) [135,140,141,143], and ZIF-8 [137,144,145].

5.4 Applications of MOFs in µSPE

MOFs have been applied for the analysis of a wide variety of samples in the different µSPE modalities. Figure 5.5 shows the different samples analyzed so far by each modality, expressed as percent of articles published in the JCR listed journals.

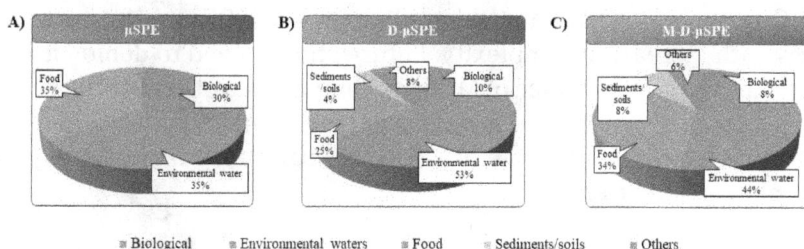

Figure 5.5 Type of samples analyzed using MOFs as sorbent materials in the different µSPE modalities.

In the static µSPE modality (Figure 5.5(A)), MOFs have been employed for the analysis of biological samples [34,36-38,44], environmental water [35,38-41,48] and food samples [42-45,47]. In spite of a limited number of samples studied so far, a wider range of analytes has been analyzed, such as aldehydes [36], drugs [34,37,38,41], flavonoids [39], hormones [47], metal ions [43,45], polycyclic aromatic hydrocarbons (PAHs) [35,48], proteins [44,46] and sulfonamides (SAs) [40].

In D-µSPE, MOFs-based materials have been tested for a wider variety of sample types (Figure 5.5(B)). In any case, the main applications reported for MOFs in D-µSPE include the determination of pollutants in environmental water [49-54,57,58,61,62,65,67,68,70,73,75,76,78,80,82-84,88,91] and food materials [59,60,64,69,72,79,84,85,88-90]. The studied pollutants include pesticides and herbicides [58-60,64,76,78], PAHs [57,63,88,91], etc., along with drugs with anti-inflammatory capacity or antibacterial activity [54,67,71,79,82,89,90].

Regarding the use of MOFs in M-D-µSPE (Figure 5.5(C)), similar sample types as D-µSPE can be pointed out: environmental water [92,93,97,98,100,101,106,107,109-111,114-116,122-131,134,136, 137,140,143,144,148], food materials [93,95-97,102-105,108,110,112,113,115,117,119,120,122,124,129-132,135,148], biological samples [99,121,128,142,145,147], sediments and soils [95-97,100,138], etc. However, the use of MOFs has been particularly focused on the determination of the heavy metal ions [93,95-97,100,102-109,112,113,115,122,130]. In these applications, it is necessary to incorporate a chelating agent to the microextraction strategy. Interestingly, several studies have reported the incorporation of the chelating agents directly into the extracting material [96,102,105,113,115,130].

Clearly, reported analytical applications of MOFs are facing samples with increasing complexity. Future studies need to demonstrate the superior performance of MOFs as compared to other commercialized sorbents materials. Particularly, when employing a composite material based on MOF, the presence of MOF as the main extraction agent needs to be ensured.

5.5 Conclusions

Metal-organic frameworks (MOFs) are microporous hybrid crystals presenting a highly ordered three-dimensional structure. They are composed of a metal (or metal cluster) and an organic linker (or more than one linker) connected by coordination bonds, which provide exceptional porosity and high surface area. Furthermore, the tuneability of MOFs (simply changing the nature of the metal and/or organic ligand) permits the synthesis of task-specific and selective materials for target applications.

In this sense, MOFs are promising materials in energy and environmental fields, gas storage, catalysis, sensing, separation processes and biological applications.

Recently, MOFs have also received attention in the analytical chemistry field as promising sorbents for the improvement of a number of extraction and pre-concentration methods. The use of MOFs as sorbents in sample preparation contributes to the development of environmental-friendly methods by removing conventional toxic organic solvents or sorbents. Most commonly, solid-phase microextraction (SPME) using fibers coatings and miniaturized (or micro-) solid-phase extraction (µSPE) without fibers coatings have been the sorbent-based microextraction methods of choice for MOFs and derivatives.

The advantages of µSPE over SPME are linked to the simplicity, as exhaustive (or almost exhaustive) extractions can be obtained in shorter durations. In µSPE, MOFs have been used (i) as bare MOFs (neat materials), (ii) in composite form with other materials and (iii) in confined form in certain devices. Independent of the microextraction mode, µSPE implies the use of lower amounts of sorbent material (<500 mg).

Given the increasing number of applications of MOFs in µSPE and the increasing number of sub-modes within µSPE, this chapter provides an overview of the topic together with a discussion on current trends.

Acknowledgment

P.G.-H. thanks the Agencia Canaria de Investigación, Innovación y Sociedad de la Información (ACIISI), co-funded by the European Social Fund, for her FPI PhD fellowship. A.G.-S. thanks CajaSiete and ULL for his FPI PhD fellowship. P.R.-B. thanks her FPI PhD research contract associated to the Project Ref. MAT2014-57465-R. J.P. thanks the "Agustín de Betancourt" Canary Program for their research associate positions at ULL. V.P. thanks the Spanish Ministry of Economy and Competitiveness (MINECO) for the Projects Ref. MAT2014-57465-R and MAT2017-89207-R.

Abbreviations

3CPS	3-chloropropyltriethoxysilane
APTES	3-aminopropyltriethoxysilane
BET	Brunauer-Emmett-Teller
CSD	Cambridge Structural Database
GAC	green analytical chemistry
GO	graphene oxide
HF	hollow fibers
HPLC	high performance liquid chromatography
IL	ionic liquid
JCR	journal of citation report
LLE	liquid-liquid extraction
MMM	mixed-matrix membranes
MOF	metal-organic framework
MS	mass spectrometry
MWCNT	multi-walled carbon nanotube
PP	polypropylene
PVA	polyvinyl alcohol
SBSE	stir-bar sorptive extraction
SBU	secondary building unit
SCSE	stir-cake sorptive microextraction
SPE	solid-phase extraction
SPME	solid-phase microextraction
TEOS	tetraethyl orthosilicate

References

1. Hoskins, B. F., and Robson, R. (1990) Design and construction of a new class of scaffolding-like materials comprising infinite polymeric frameworks of 3D-linked molecular rods. A reappraisal of the zinc cyanide and cadmium cyanide structures and the synthesis and structure of the diamond-related frameworks [N(CH$_3$)$_4$][Cu^{I}Zn^{II}(CN)$_4$] and Cu^I[4,4',4'',4'''-tetracyanotetraphenylmethane]BF$_4$.xC$_6$H$_5$NO$_2$. *Journal of the American Chemical Society*, **112**, 1546-1554.

2. Kitagawa, S., Matsuyama, S., Munakata, M., and Emori, T. (1991) Synthesis and crystal structures of novel one-dimensional polymers, [{M(bpen)X}∞][M = Cu^I, X = PF$_6$⁻; M = Ag^I, X = ClO$_4$⁻; bpen =trans-1,2-bis(2-pyridyl)ethylene] and [{Cu(bpen)(CO)(CH$_3$CN)(PF$_6$)}∞]. *Journal of the Chemical Society, Dalton Transactions*, 2869-2874.

3. Yaghi, O. M., and Li, H. (1995) Hydrothermal synthesis of a metal-organic framework containing large rectangular channels. *Journal of the American Chemical Society*, **117**, 10401-10402.

4. Riou, D., and Férey, G. (1998) Hybrid open frameworks (MIL-n). Part 3 Crystal structures of the HT and LT forms of MIL-7: a new vanadium propylenediphosphonate with an open-framework. Influence of the synthesis temperature on the oxidation state of vanadium within the same structural type. *Journal of Materials Chemistry*, **8**, 2733-2735.

5. Yaghi, O. M. (2016) Reticular chemistry-construction, properties, and precision reactions of frameworks. *Journal of the American Chemical Society*, **138**, 15507-15509.

6. Furukawa, H., Cordova, K. E., O'Keeffe, M., and Yaghi, O. M. (2013) The chemistry and applications of metal-organic framework. *Science*, **341**, 1230444.

7. Kumar, P., Vellingiri, K.; Kim, K. H., Brown, R. J. C., and Manos, M. J. (2017) Modern progress in metal-organic frameworks and their composites for diverse applications. *Microporous and Mesoporous Materials*, **253**, 251-265.

8. Pettinari, C., Marchetti, F., Mosca, N., Tosi, G., and Drozdov, A. (2017) Application of metal–organic frameworks. *Polymer International*, **66**, 731-744.

9. Ma, S., and Zhou, H.-C. (2010) Gas storage in porous metal–organic frameworks for clean energy applications. *Chemical Communications*, **46**, 44-53.

10. Yang, X., and Xu, Q. (2017) Bimetallic metal-organic frameworks for gas storage and separation. *Crystal Growth & Design*, **17**, 1450-1455.

11. Wu, M.-X., and Yang, Y.-W. (2017) Metal–organic framework (MOF)-based drug/cargo delivery and cancer therapy. *Advanced Materials*,

29, 1606134.

12. Della Rocca, J., Liu, D., and Lin, W. (2011) Nanoscale metal-organic frameworks for biomedical imaging and drug delivery. *Accounts of Chemical Research*, **44**, 957-968.

13. Zhu, L., Liu, X.-Q., Jiang, H.-L., and Sun, L.-B. (2017) Metal–organic frameworks for heterogeneous basic catalysis. *Chemical Reviews*, **117**, 8129-8176.

14. Ranocchiari, M., and van Bokhoven, J. A. (2011) Catalysis by metal organic frameworks: fundamentals and opportunities. *Physical Chemistry Chemical Physics*, **13**, 6388-6396.

15. Kumar, P., Deep, A., and Kim, K.-H. (2015) Metal organic frameworks for sensing applications. *Trends in Analytical Chemistry*, **73**, 39-53.

16. Lustig, W. P., Mukherjee, S., Rudd, N. D., Desai, A. V., Li, J., and Ghosh, S. K. (2017) Metal-organic frameworks: functional luminescent and photonic materials for sensing applications. *Chemical Society Reviews*, **46**, 3242-3285.

17. McKinlay, A. C., Morris, R. E., Horcajada, P., Ferey, G., Gref, R., Couvreur, P., and Serre, C. (2010) BioMOFs: Metal-organic frameworks for biological and medical applications. *Angewandte Chemie International Edition*, **49**, 6260-6266.

18. Cui, Y., Zhu, F., Chen, B., and Qian, G. (2015) Metal–organic frameworks for luminescence thermometry. *Chemical Communications*, **51**, 7420-7431.

19. Rocío-Bautista, P., González-Hernández, P., Pino, V., Pasán, J., and Afonso, A. M. (2017) Metal-organic frameworks as novel sorbents in dispersive-based microextraction approaches. *Trends in Analytical Chemistry*, **90**, 114-134.

20. Rocío-Bautista, P., Pacheco-Fernández, I., Pasán, J., Pino, V. (2016) Are metal-organic frameworks able to provide a new generation of solid-phase microextraction coatings? - A review. *Analytica Chimica Acta*, **939**, 26-41.

21. Maya, F., Cabello, C. P., Frizzarin, R. M., Estela, J. M., Palomino, G. T., and Cerdà, V. (2017) Magnetic solid-phase extraction using metal-organic frameworks (MOFs) and their derived carbons. *Trends in Analytical Chemistry*, **90**, 142-152.

22. Zhang, J., and Chen, Z. (2017) Metal-organic frameworks as stationary phase for application in chromatographic separation. *Journal of Chromatography A*, **1530**, 1-18.

23. Yusuf, K., Aqel, A., and Alothman, Z. (2014) Metal-organic frameworks in chromatography. *Journal of Chromatography A*, **1348**, 1-16.

24. Pacheco-Fernández, I., González-Hernández, P., Pasán, J., Ayala, J. H., and Pino, V. (2018) The rise of metal-organic frameworks in analytical chemistry. In: *Handbook of Smart Materials in Analytical Chem-*

istry, De La Guardia, M., and Esteve-Trurrillas, F. A. (eds), volume 1, Wiley, Germany, *in print.*

25. Hashemia, B., Zohrabi, P., and Shamsipur, M., (2018) Recent developments and applications of different sorbents for SPE and SPME from biological samples. *Talanta*, **187**, 337-347.

26. Pawliszyn, J. (2000) Theory of Solid-Phase Microextraction. *Journal of Chromatographic Science*, **38**, 270-278.

27. Piri-Moghadam, H., Alam, M. N., and Pawliszyn, J. (2017) Review of geometries and coating materials in solid phase microextraction: Opportunities, limitations, and future perspectives. *Analytica Chimica Acta*, **984**, 42-65.

28. Tian, J., Xu, J., Zhu, F., Lu, T., Su, C., and Ouyang, G. (2013) Application of nanomaterials in sample preparation. *Journal of Chromatography A*, **1300**, 2-16.

29. Wen, Y., Chen, L., Li, J., Liu, D., and Chen, L. (2014) Recent advances in solid-phase sorbents for sample preparation prior to chromatographic analysis. *Trends in Analytical Chemistry*, **59**, 26-41.

30. *SPME Fiber Assemblies*, Sigma Aldrich (2018). Online: https://www.sigmaaldrich.com/analytical-chromatography/analytical-products.html?TablePage=9645337 [accessed 10th July 2018].

31. *Restek PAL SPME Fibers*, Restek (2018). Online: http://www.restek.com/catalog/view/47352 [accessed 10th July 2018].

32. Pacheco-Fernández, I., Gutiérrez-Serpa, A., Rocío-Bautista, P., and Pino, V. (2017) *Molecularly Imprinted Polymers as Promising Sorbents in SPME Applications*, Nova Science Publishers, USA.

33. Lucena, R., Simonet, B. M., Cárdenas, S., and Valcárcel, M. (2011) Potential of nanoparticles in sample preparation. *Journal of Chromatography A*, **1218**, 620-637.

34. Hu, Y., Song, C., Liao, J., Huang, Z., and Li, G. (2013) Water stable metal-organic framework packed microcolumn for online sorptive extraction and direct analysis of naproxen and its metabolite from urine sample. *Journal of Chromatography A*, **1294**, 17-24.

35. Yang, S., Chen, C., Yan, Z., Cai, Q., and Yao, S. (2013) Evaluation of metal-organic framework 5 as a new SPE material for the determination of polycyclic aromatic hydrocarbons in environmental waters. *Journal of Separation Science*, **36**, 1283-1290.

36. Wang, S., Wang, X., Ren, Y., and Xu, H. (2015) Metal-organic framework 199 film as a novel adsorbent of thin-film extraction. *Chromatographia*, **78**, 621-629.

37. Asiabi, M., Mehdinia, A., and Jabbari, A. (2015) Preparation of water stable methyl-modified metal–organic framework-5/polyacrylonitrile composite nanofibers via electrospinning and their application for solid-phase extraction of two estrogenic drugs in urine samples.

Journal of Chromatography A, **1426**, 24-32.

38. Lyu, D.-Y., Yang, C.-X., and Yan, X.-P. (2015) Fabrication of aluminum terephthalate metal-organic framework incorporated polymer monolith for the microextraction of non-steroidal anti-inflammatory drugs in water and urine samples. *Journal of Chromatography A*, **1393**, 1-17.

39. Liu, Y., Hu, J., Li, Y., Li, X.-S., and Wang, Z.-L. (2016) Metal-organic framework MIL-101 as sorbent based on double-pumps controlled on-line solid-phase extraction coupled with high-performance liquid chromatography for the determination of flavonoids in environmental water samples. *Electrophoresis*, **37**, 2478-2486.

40. Dai, X., Jia, X., Zhao, P., Wang, T., Wang, J., Huang, P., He, L., and Hou, X. (2016) A combined experimental/computational study on metal-organic framework MIL-101(Cr) as a SPE sorbent for the determination of sulphonamides in environmental water samples coupling with UPLC-MS/MS. *Talanta*, **154**, 581-588.

41. Shih, Y.-H., Kuo, Y.-C., Lirio, S., Wang, K.-Y., Lin, C.-H., and Huang, H.-Y. (2017) A simple approach to enhance the water stability of a metal-organic framework. *Chemistry: A European Journal*, **23**, 42-46.

42. Li, D., Zhang, X., Kong, F., Qiao, X., and Xu, Z. (2017) Molecularly imprinted solid-phase extraction coupled with high-performance liquid chromatography for the determination of trace trichlorfon and monocrotophos residues in fruits. *Food Analytical Methods*, **10**, 1284-1292.

43. Kahkha, M. R. R., Daliran, S., Oveisi, A. R., Kaykhaii, M., and Sepehri, Z. (2017) The mesoporous porphyrinic zirconium metal-organic framework for pipette-tip solid-phase extraction of mercury from fish samples followed by cold vapor atomic absorption spectrometric determination. *Food Analytical Methods*, **10**, 2175-2184.

44. Li, D., Yin, D., Chen, Y., and Liu, Z. (2017) Coupling of metal-organic frameworks-containing monolithic capillary-based selective enrichment with matrix-assisted laser desorption ionization-time-of-flight mass spectrometry for efficient analysis of protein phosphorylation. *Journal of Chromatography A*, **1498**, 56-63.

45. Asiabi, M., Mehdinia, A., and Jabbari, A. (2017) Spider-web-like chitosan/MIL-68(Al) composite nanofibers for high-efficient solid phase extraction of Pb(II) and Cd(II). *Microchimica Acta*, **184**, 4495-4501.

46. Li, D., and Bie, Z. (2017) Metal–organic framework incorporated monolithic capillary for selective enrichment of phosphopeptides. *RSC Advances*, **7**, 15894-15902.

47. Yan, Z., Wu, M., Hu, B., Yao, M., Zhang, L., Lu, Q., and Pang, J. (2018) Electrospun UiO-66/polyacrylonitrile nanofibers as efficient sorbent for pipette tip solid phase extraction of phytohormones in

vegetable samples. *Journal of Chromatography A*, **1542**, 19-27.

48. Zhang, X., Wang, P., Han, Q., Li, H., Wang, T., and Ding M. (2018) Metal–organic framework based in-syringe solid-phase extraction for the on-site sampling of polycyclic aromatic hydrocarbons from environmental water samples. *Journal of Separation Science*, **41**, 1856-1863.

49. Ge, D., and Lee, H. K. (2011) Water stability of zeolite imidazolate framework 8 and application to porous membrane-protected micro-solid-phase extraction of polycyclic aromatic hydrocarbons from environmental water samples. *Journal of Chromatography A*, **1218**, 8490-8495.

50. Ge, D., and Lee, H. K. (2012) Sonication-assisted emulsification microextraction combined with vortex-assisted porous membrane-protected micro-solid-phase extraction using mixed zeolitic imidazolate frameworks 8 as sorbent. *Journal of Chromatography A*, **1263**, 1-6.

51. Zang, H., Yuan, J.-P., Chen, X.-F., Liu, C.-A., Cheng, C.-G., and Zhao, R.-S. (2013) Hollow fiber-protected metal–organic framework materials as micro-solid-phase extraction adsorbents for the determination of polychlorinated biphenyls in water samples by gas chromatography-tandem mass spectrometry. *Analytical Methods*, **5**, 4875-4882.

52. Wang, T., Wang, J., Zhang, C., Yang, Z., Dai, X., Cheng, M., and Hou, X. (2015) Metal–organic framework MIL-101(Cr) as a sorbent of porous membrane-protected micro-solid-phase extraction for the analysis of six phthalate esters from drinking water: A combination of experimental and computational study. *Analyst*, **140**, 5308-5316.

53. Huang, Z., and Lee, H. K. (2015) Micro-solid-phase extraction of organochlorine pesticides using porous metal-organic framework MIL-101 as sorbent. *Journal of Chromatography A*, **1401**, 9-16.

54. Wang, Y., Zhang, Y., Cui, J., Li, S., Yuan, M., Wang, T., Hu, Q., and Hou, X. (2018) Fabrication and characterization of metal organic frameworks/polyvinyl alcohol cryogel and their application in extraction of non-steroidal anti-inflammatory drugs in water samples. *Analytica Chimica Acta*, **1022**, 45-52.

55. Gao, G., Li, S., Li, S., Zhao, L., Wang, T., and Hou, X. (2018) Development and application of vortex-assisted membrane extraction based on metal-organic framework mixed-matrix membrane for the analysis of estrogens in human urine. *Analytica Chimica Acta*, **1023**, 35-43.

56. Zhai, Y., Li, N., Lei, L., Yang, X., and Zhang, H. (2014) Dispersive micro-solid-phase extraction of hormones in liquid cosmetics with metal–organic framework. *Analytical Methods*, **6**, 9435-9445.

57. Hu, H., Liu, S., Chen, C., Wang, J., Zou, Y., Lin, L., and Yao, S. (2014) Two novel zeolitic imidazolate frameworks (ZIFs) as sorbents for

solid-phase extraction (SPE) of polycyclic aromatic hydrocarbons (PAHs) in environmental water samples. *Analyst*, **139**, 5818-5826.

58. Li, X., Xing, J., Chang, C., Wang, X., Bai, Y., Yan, X., and Liu, H. (2014) Solid-phase extraction with the metal–organic framework MIL-101(Cr) combined with direct analysis in real time mass spectrometry for the fast analysis of triazine herbicides. *Journal of Separation Science*, **37**, 1489-1495.

59. Zhibing, N. L., Wang, Z., Zhang, L., Nian, L., Lei, L., Yang, X., Zhang, H., and Yu, A. (2014) Liquid-phase extraction coupled with metal–organic frameworks-based dispersive solid phase extraction of herbicides in peanuts. *Talanta*, **128**, 345-353.

60. Li, N., Zhang, L., Nian, L., Cao, B., Wang, Z., Lei, L., Yang, X., Sui, J., Zhang, H., and Yu, A. (2015) Dispersive micro-solid-phase extraction of herbicides in vegetable oil with metal–organic framework MIL-101. *Journal of Agricultural and Food Chemistry*, **63**, 2154-2161.

61. Tahmasebi, E., Masoomi, M. Y., Yamini, Y., and Morsali, A. (2015) Application of mechanosynthesized azine-decorated zinc(II) metal–organic frameworks for highly efficient removal and extraction of some heavy-metal ions from aqueous samples: A comparative study. *Inorganic Chemistry*, **54**, 425-433.

62. Su, H., Wang, Z., Jia, Y., Deng, L., Chen, X., Zhao, R., and Chan, T.-W. D. (2015) A cadmium(II)-based metal-organic framework material for the dispersive solid-phase extraction of polybrominated diphenyl ethers in environmental water samples. *Journal of Chromatography A*, **1422**, 334-339.

63. Rocío-Bautista, P., Martínez-Benito, C., Pino, V., Pasán, J., Ayala, J. H., Ruíz-Pérez, C., and Afonso, A. M. (2015) The metal–organic framework HKUST-1 as efficient sorbent in a vortex-assisted dispersive micro solid-phase extraction of parabens from environmental waters, cosmetic creams, and human urine. *Talanta*, **139**, 13-20.

64. Li, N., Wu, L., Song, Y., Lei, L., Yang, X., Wang, K., Wang, Z., Zhang, L., Zhang, H., Yu, A., and Zhang, Z. (2015) Dynamic microwave assisted extraction coupled with dispersive micro-solid-phase extraction of herbicides in soybeans. *Talanta*, **142**, 43-50.

65. Huang, Z., and Lee, H. K. (2015) Performance of metal-organic framework MIL-101 after surfactant modification in the extraction of endocrine disrupting chemicals from environmental water samples. *Talanta*, **143**, 366-373.

66. Li, N., Zhu, Q., Yang, Y., Huang, J., Dang, X., and Chen, H. (2015) A novel dispersive solid-phase extraction method using metal-organic framework MIL-101 as the adsorbent for the analysis of benzophenones in toner. *Talanta*, **132**, 713-718.

67. Lu, N., Wang, T., Zhao, P., Zhang, L., Lun, X., Zhang, X., and Hou, X. (2016) Experimental and molecular docking investigation on

metal-organic framework MIL-101(Cr) as a sorbent for vortex assisted dispersive micro-solid-phase extraction of trace 5-nitroimidazole residues in environmental water samples prior to UPLC-MS/MS analysis. *Analytical and Bioanalytical Chemistry*, **408**, 8515-8528.

68. Jamali, A., Tehrani, A. A., Shemirani, F., and Morsali, A. (2016) Lanthanide metal-organic frameworks as selective microporous materials for adsorption of heavy metal ions. *Dalton Transactions*, **45**, 9193-9200.

69. Wu, Y., Xu, G., Wei, F., Song, Q., Tang, T., Wang, X., and Hu, Q. (2016) Determination of Hg (II) in tea and mushroom samples based on metal-organic frameworks as solid phase extraction sorbents. *Microporous and Mesoporous Materials*, **235**, 204-210.

70. Tahmasebi, E., Yaser, M., Masoomi, Y., Yamini, Y., and Morsali, A. (2016) Application of a Zn(II) based metal–organic framework as an efficient solid-phase extraction sorbent for preconcentration of plasticizer compounds. *RSC Advances*, **6**, 40211-40218.

71. Qu, F., Xia, L., Wu, C., Liu, L., Li, G., and You, J. (2016) Sensitive and accurate determination of sialic acids in serum with the aid of dispersive solid-phase extraction using the zirconium-based MOF of UiO-66-NH$_2$ as sorbent. *RSC Advances*, **6**, 64895-64901.

72. Liu, L., Xia, L., Wu, C., Qu, F., Li, G., Sun, Z., and You, J. (2016) Zirconium (IV)-based metal organic framework (UIO-67) as efficient sorbent in dispersive solid phase extraction of plant growth regulator from fruits coupled with HPLC fluorescence detection. *Talanta*, **154**, 23-30.

73. Lv, Z., Sun, Z., Song, C., Lu, S., Chen, G., and You, J. (2016) Sensitive and background-free determination of thiols from wastewater samples by MOF-5 extraction coupled with high-performance liquid chromatography with fluorescence detection using a novel fluorescence probe of carbazole-9-ethyl-2-maleimide. *Talanta*, **161**, 228-237.

74. Tokalioglu, S., Yavuz, E., Demir, S., and Patat, S. (2017) Zirconium-based highly porous metal-organic framework (MOF-545) as an efficient adsorbent for vortex assisted-solid phase extraction of lead from cereal, beverage and water samples. *Food Chemistry*, **237**, 707-715.

75. Cai, Q., Zhang, L., Zhao, P., Lun, X., Li, W., Guo, Y., and Hou, W. (2017) A joint experimental-computational investigation: Metal organic framework as a vortex assisted dispersive micro-solid-phase extraction sorbent coupled with UPLC-MS/MS for the simultaneous determination of amphenicols and their metabolite in aquaculture water. *Microchemical Journal*, **130**, 263-270.

76. Wang, X., Ma, X., Wang, H., Huang, P., Du, X., and Lu, X. (2017) A zinc(II) benzenetricarboxylate metal organic framework with unu-

sual adsorption properties, and its application to the preconcentration of pesticides. *Microchimica Acta*, **184**, 3681-3687.

77. Cui, Y.-Y., Yang, C.-X., Yang, X.-D., and Yan, X.,-P. (2018) Zeolitic imidazolate framework-8 for selective extraction of a highly active anti-oxidant flavonoid from Caragana Jubata. *Journal of Chromatography A*, **1544**, 8-15.

78. Cao, X., Jiang, Z., Wang, S., Hong, S., Li, H., Zhang, C., Shao, Y., She, Y., Jin, F., Jin, M., and Wang, J. Metal-organic framework UiO-66 for rapid dispersive solid phase extraction of neonicotinoid insecticides in water samples. *Journal of Chromatography B*, **1077-1078**, 92-97.

79. Li, Y., Zhu, N., Chen, T., Ma, Y., and Li, Q. (2018) A green cyclodextrin metal-organic framework as solid-phase extraction medium for enrichment of sulfonamides before their HPLC determination. *Microchemical Journal*, **138**, 401-407.

80. Moghaddam, Z. S., Kaykhaii, M., Khajeh, M., and Oveisi, A. R. (2018) Synthesis of UiO-66-OH zirconium metal-organic framework and its application for selective extraction and trace determination of thorium in water samples by spectrophotometry. *Spectrochimica Acta Part A: Molecular and Biomolecular Spectroscopy*, **194**, 76-82.

81. Gao, G., Li, S., Li, S., Wang, Y., Zhao, P., Zhang, X., and Hou, X. (2018) A combination of computational–experimental study on metal-organic frameworks MIL-53(Al) as sorbent for simultaneous determination of estrogens and glucocorticoids in water and urine samples by dispersive micro-solid-phase extraction coupled to UPLC-MS/MS. *Talanta*, **180**, 358-367.

82. Rocío-Bautista, P., Pino, V., Pasán, J., López-Hernández, I., Ayala, J. H., Ruíz-Pérez, C., Afonso, A. M. (2018) Insights in the analytical performance of neat metal-organic frameworks in the determination of pollutants of different nature from waters using dispersive miniaturized solid-phase extraction and liquid chromatography. *Talanta*, **179**, 775-783.

83. Ke, F., Qiu, L.-G., Yuan, Y.-P., Peng, F.-M., Jiang, X., Xie, A.-J., Shen, Y.-H., and Xhu, J.-F. (2011) Thiol-functionalization of metal-organic framework by a facile coordination-based postsynthetic strategy and enhanced removal of Hg^{2+} from water. *Journal of Hazardous Materials*, **196**, 36-43.

84. Sohrabi, M. R. (2014) Preconcentration of mercury(II) using a thiol-functionalized metal-organic framework nanocomposite as a sorbent. *Microchimica Acta*, **181**, 435-444.

85. Wang, Y., Wu, Y., Ge, H., Chen, H., Ye, G., and Hu, X. (2014) Fabrication of metal-organic frameworks and graphite oxide hybrid composites for solid-phase extraction and preconcentration of luteolin. *Talanta*, **122**, 91-96.

86. Zhang, C.-Y., Yan, Z.-G., Zhou, Y.-Y., Wang, L., Xie, Y.-B., Bai, L.-P.,

Zhou, H.-Y., and Li, F.-S. (2015) Embedment of Ag(I)-organic frameworks into silica gels for microextraction of polybrominated diphenyl ethers in soils. *Journal of Chromatography A*, **1383**, 18-24.

87. Qiang, Y., Wang, W.-F., Dhodary, B., and Yang, J.-L. (2017) Zeolitic imidazolate framework 8 (ZIF-8) reinforced macroporous resin D101 for selective solid-phase extraction of 1-naphthol and 2-naphthol from phenol compounds. *Electrophoresis*, **38**, 1685-1692.

88. Nasrollahpour, A., Moradi, S. E., and Baniamerian, M. J. (2017) Vortex-assisted dispersive solid-phase microextraction using ionic liquid-modified metal-organic frameworks of PAHs from environmental water, vegetable, and fruit juice samples. *Food Analytical Methods*, **10**, 2815-2826.

89. Wang, Y., Dai, X., He, X., Chen, L., and Hou, X. (2017) MIL-101(Cr)@GO for dispersive micro-solid-phase extraction of pharmaceutical residue in chicken breast used in microwave-assisted coupling with HPLC–MS/MS detection. *Journal of Pharmaceutical and Biomedical Analysis*, **145**, 440-446.

90. Jia, X., Zhao, P., Ye, X., Zhang, L., Wang, T., Chen, Q., and Hou, X. (2017) A novel metal-organic framework composite MIL-101(Cr)@GO as an efficient sorbent in dispersive micro-solid phase extraction coupling with UHPLC-MS/MS for the determination of sulfonamides in milk samples. *Talanta*, **169**, 227-238.

91. Jia, Y., Zhao, Y., Zhao, M., Wang, Z., Chen, X., and Wang, M. (2018) Core–shell indium (III) sulfide@metal-organic framework nanocomposite as an adsorbent for the dispersive solid-phase extraction of nitro-polycyclic aromatic hydrocarbons. *Journal of Chromatography A*, **1551**, 21-28.

92. Huo, S.-H., and Yan, X.-P (2012) Facile magnetization of metal–organic framework MIL-101 for magnetic solid-phase extraction of polycyclic aromatic hydrocarbons in environmental water samples. *Analyst*, **137**, 3445-3451.

93. Bagheri, A., Taghizadeh, M., Behbahani, M., Asgharinezhad, A. A., Salarian, M., Dehghani, A., Ebrahimzadeh, H., and Amini, M. M. (2012) Synthesis and characterization of magnetic metal-organic framework (MOF) as a novel sorbent, and its optimization by experimental design methodology for determination of palladium in environmental samples. *Talanta*, **99**, 132-139.

94. Hu, Y., Huang, Z., Liao, J., and Li, G. (2013) Chemical bonding approach for fabrication of hybrid magnetic metal–organic framework-5: High efficient adsorbents for magnetic enrichment of trace analytes. *Analytical Chemistry*, **85**, 6885-6893.

95. Wang, Y., Xie, J., Wu, Y., Ge, H., and Hu, X. (2013) Preparation of a functionalized magnetic metal–organic framework sorbent for the extraction of lead prior to electrothermal atomic absorption spectrometer analysis. *Journal of Materials Chemistry A*, **1**, 8782-8789.

96. Taghizadeh, M., Asgharinezhad, A. A., Pooladi, M., Barzin, M., Abbaszadeh, A., and Tadjarodi, A. (2013) A novel magnetic metal organic framework nanocomposite for extraction and preconcentration of heavy metal ions, and its optimization via experimental design methodology. *Microchimica Acta*, **180**, 1073-1084.

97. Sohrabi, M. R., Matbouie, Z., Asgharinezhad, A. A., and Dehghani, A. (2013) Solid phase extraction of Cd (II) and Pb (II) using a magnetic metal-organic framework, and their determination by FAAS. *Microchimica Acta*, **180**, 589-597.

98. Wang, G.-H., Lei, Y.-Q., and Song, H.-C. (2014) Evaluation of $Fe_3O_4@SiO_2$–MOF-177 as an advantageous adsorbent for magnetic solid-phase extraction of phenols in environmental water samples. *Analytical Methods*, **6**, 7842-7847.

99. Zhang, S., Jiao, Z., and Yao, W. (2014) A simple solvothermal process for fabrication of a metal-organic framework with an iron oxide enclosure for the determination of organophosphorus pesticides in biological samples. *Journal of Chromatography A*, **1371**, 74-81.

100. Wang, Y., Xie, J., Wu, Y., and Hu, X. (2014) A magnetic metal-organic framework as a new sorbent for solid-phase extraction of copper (II), and its determination by electrothermal AAS. *Microchimica Acta*, **181**, 949-956.

101. Maya, F., Cabello, C. P., Estela, J. M., Cerdà, V., and Palomino, G. T. (2015) Automatic in-syringe dispersive microsolid phase extraction using magnetic metal–organic frameworks. *Analytical Chemistry*, **87**, 7545-7549.

102. Ghorbani-Kalhor, E:, Hosseinzadeh-Khanmiri, R., Babazadeh, M., Abolhasani, J., and Hassanpour, A. (2014) Synthesis and application of a novel magnetic metal-organic framework nanocomposite for determination of Cd, Pb, and Zn in baby food samples. *Canadian Journal of Chemistry*, **93**(5), 518-525.

103. Hassanpour, A., Hosseinzadeh Khanmiri, R., Babazadeh, M., Abolhasani, J, and Ghorbani-Kalhor, E. (2015) Determination of heavy metal ions in vegetable samples using a magnetic metal-organic framework nanocomposite sorbent. *Food Additives & Contaminants, Part A*, **32**(5), 725-736.

104. Wang, Y., Chen, H., Tang, J., Ye, G., Ge, H., and Hu, X. (2015) Preparation of magnetic metal organic frameworks adsorbent modified with mercapto groups for the extraction and analysis of lead in food samples by flame atomic absorption spectrometry. *Food Chemistry*, **181**, 191-197.

105. Ghorbani-Kalhor, E., Hosseinzadeh–Khanmiri, R., Abolhasani, J., Babazadeh, M., and Hassanpour, A. (2015) Determination of mercury(II) ions in seafood samples after extraction and preconcentration by a novel functionalized magnetic metal–organic framework nanocomposite. *Journal of Separation Science*, **38**, 1179-1186.

106. Xu, Y., Jian, J., Li, X., Han, Y., Meng, H., Song, C., and Zhang, X. (2015) Magnetization of a Cu(II)-1,3,5-benzenetricarboxylate metal-organic framework for efficient solid-phase extraction of Congo Red. *Microchimica Acta*, **182**, 2313-2320.

107. Ricco, R., Konstas, K., Styles, M. J., Richardson, J. J., Babarao, R., Suzuki, K., Scopece, P., and Falcaro, P. (2015) Lead (II) uptake by aluminium based magnetic framework composites (MFCs) in water. *Journal of Materials Chemistry A*, **3**, 19822-19831.

108. Babazadeh, M., Hosseinzadeh-Khanmiri, R., Abolhasani, J., Ghorbani-Kalhor, E., and Hassanpour, A. (2015) Solid phase extraction of heavy metal ions from agricultural samples with the aid of a novel functionalized magnetic metal–organic framework. *RSC Advances*, **5**, 19884-19892.

109. Moradi, S. E., Shabani, A. M. H., Dadfamia, S., and Emami, S. (2016) Sulfonated metal organic framework loaded on iron oxide nanoparticles as a new sorbent for the magnetic solid phase extraction of cadmium from environmental water samples. *Analytical Methods*, **8**, 6337-6346.

110. Rocío-Bautista, P., Pino, V., Ayala, J. H., Pasán, J., Ruiz-Pérez, C., and Afonso, A. M. (2016) A magnetic-based dispersive micro-solid-phase extraction method using the metal-organic framework HKUST-1 and ultra-high-performance liquid chromatography with fluorescence detection for determining polycyclic aromatic hydrocarbons in waters and fruit tea infusions. *Journal of Chromatography A*, **1436**, 42-50.

111. Zhang, S., Yao, W., Ying, J., and Zhao, H. (2016) Polydopamine-reinforced magnetization of zeolitic imidazolate framework ZIF-7 for magnetic solid-phase extraction of polycyclic aromatic hydrocarbons from the air-water environment. *Journal of Chromatography A*, **1452**, 18-26.

112. Tadjarodi, A., and Abbaszadeh, A. (2016) A magnetic nanocomposite prepared from chelator-modified magnetite (Fe_3O_4) and HKUST-1 (MOF-199) for separation and preconcentration of mercury (II). *Microchimica Acta*, **183**, 1391-1399.

113. Ghorbani-Kalhor, E. (2016) A metal-organic framework nanocomposite made from functionalized magnetite nanoparticles and HKUST-1 (MOF-199) for preconcentration of Cd (II), Pb (II), and Ni (II). *Microchimica Acta*, **183**, 2639-2647.

114. Liu, H., Ren, X., and Chen, L. (2016) Synthesis and characterization of magnetic metal–organic framework for the adsorptive removal of Rhodamine B from aqueous solution. *Journal of Industrial and Engineering Chemistry*, **34**, 278-285.

115. Abbaszadeh, A., and Tadjarodi, A. (2016) Speciation analysis of inorganic arsenic in food and water samples by electrothermal atomic absorption spectrometry after magnetic solid phase extraction by a

novel MOF-199/modified magnetite nanoparticle composite. *RSC Advances*, **6**, 113727-113736.

116. Ma, J., Yao, Z., Hou, L., Lu, W., Yang, Q., Li, J., and Chen, L. (2016) Metal organic frameworks (MOFs) for magnetic solid-phase extraction of pyrazole/pyrrole pesticides in environmental water samples followed by HPLC-DAD determination. *Talanta*, **161**, 686-692.

117. Li, Z., Qi, M., Tu, C., Wang, W., Chen, J., and Wang, A.-J. (2017) Magnetic metal-organic framework/graphene oxide-based solid-phase extraction combined with spectrofluorimetry for the determination of enrofloxacin in milk sample. *Food Analytical Methods*, **10**, 4094-4103.

118. Zhang, S., Han, P., and Xia, Y. (2017) Facile extraction of azide in sartan drugs using magnetized anion-exchange metal-organic frameworks prior to ion chromatography. *Journal of Chromatography A*, **1514**, 29-35.

119. Xia, L., Liu, L., Lv, X., Qu, F., Li, G., and You, J. (2017) Towards the determination of sulfonamides in meat samples: A magnetic and mesoporous metal-organic framework as an efficient sorbent for magnetic solid phase extraction combined with high-performance liquid chromatography. *Journal of Chromatography A*, **1500**, 24-31.

120. Li, W.-K., Zhang, H.-X., and Shi, Y.-P. (2017) Selective determination of aromatic amino acids by magnetic hydroxylated MWCNTs and MOFs based composite. *Journal of Chromatography B*, **1059**, 27-34.

121. Bahrani, S., Ghaedi, M., Dashtian, K., Ostovan, A., Mansoorkhani, M. J. K., and Salehi, A. MOF-5(Zn)-Fe2O4 nanocomposite based magnetic solid-phase microextraction followed by HPLC-UV for efficient enrichment of colchicine in root of colchicium extracts and plasma samples. *Journal of Chromatography B*, **1067**, 45-52.

122. Saboori, A. (2017) A nanoparticle sorbent composed of MIL-101(Fe) and dithiocarbamate-modified magnetite nanoparticles for speciation of Cr(III) and Cr(VI) prior to their determination by electrothermal AAS. *Microchimica Acta*, **184**, 1509-1516.

123. Wang, T., Liu, S., Gao, G., Zhao, P., Lu, N., Lun, X., and Hou, X. (2017) Magnetic solid phase extraction of non-steroidal anti-inflammatory drugs from water samples using a metal organic framework of type Fe_3O_4/MIL-101(Cr), and their quantitation by UPLC-MS/MS. *Microchimica Acta*, **184**, 2981-2990.

124. Safari, M., Yamini, Y., Masoomi, M. Y., Morsali, Mani-Varnosfaderami, A. (2017) Magnetic metal-organic frameworks for the extraction of trace amounts of heavy metal ions prior to their determination by ICP-AES. *Microchimica Acta*, **184**, 1555-1564.

125. Zhou, Q., Lei, M., Liu, Y., Zhao, K., and Zhao, D. (2017) Magnetic solid phase extraction of N- and S-containing polycyclic aromatic hydrocarbons at ppb levels by using a zerovalent iron nanoscale material modified with a metal organic framework of type Fe@MOF-5, and

their determination by HPLC. *Microchimica Acta*, **184**, 1029-1036.

126. Ma, J., Wu, G., Li, S., Tan, W., Wang, X., Li, J., and Chen, L. (2018) Magnetic solid-phase extraction of heterocyclic pesticides in environmental water samples using metal-organic frameworks coupled to high performance liquid chromatography determination. *Journal of Chromatography A*, **1553**, 57-66.

127. Huang, Y.-F., Liu, Q.-H., Li, K., Li, Y., and Chang, N. (2018) Magnetic iron (III)-based framework composites for the magnetic solid-phase extraction of fungicides from environmental water samples. *Journal of Separation Science*, **41**(5), 1129-1137.

128. Dargahi, .R., Ebrahimzadeh, H., Asgharinezhad, A. A., Hashemzadeh, A., Amini, M. M. (2018) Dispersive magnetic solid-phase extraction of phthalate esters from water samples and human plasma based on a nanosorbent composed of MIL-101(Cr) metal–organic framework and magnetite nanoparticles before their determination by GC–MS. *Journal of Separation Science*, **41**(4), 948-957.

129. Lu, N., He, X., Wang, T., Liu, S., and Hou, X. (2018) Magnetic solid-phase extraction using MIL-101(Cr)-based composite combined with dispersive liquid-liquid microextraction based on solidification of a floating organic droplet for the determination of pyrethroids in environmental water and tea samples. *Microchemical Journal*, **137**, 449-495.

130. Kalantari, H., and Manoochehri, M. (2018) A nanocomposite consisting of MIL-101(Cr) and functionalized magnetite nanoparticles for extraction and determination of selenium(IV) and selenium(VI). *Microchimica Acta*, **185**, 196.

131. Wang, X., Ma, X., Huang, P., Wang, J., Du, T., Du, X., and Lu, X. (2018) Magnetic Cu-MOFs embedded within graphene oxide nanocomposites for enhanced preconcentration of benzenoid-containing insecticides. *Talanta*, **181**, 112-117.

132. Liang, L., Wang, X., Sun, Y., Ma, P., Li, X., Piao, H:, Jiang, Y., and Song, D. (2018) Magnetic solid-phase extraction of triazine herbicides from rice using metalorganic framework MIL-101(Cr) functionalized magnetic particles. *Talanta*, **179**, 512-519.

133. Nooreini, M. G., and Panahi, H. A. (2016) Fabrication of magnetite nano particles and modification with metal organic framework of Zn^{2+} for sorption of doxycycline. *International Journal of Pharmaceutics*, **512**, 178-185.

134. Chen, X., Ding, N., Zang, H., Yeung, H., Zhao, R.-S., Cheng, C., Liu, J., and Chan, T.-W. D. (2013) Fe_3O_4@MOF core–shell magnetic microspheres for magnetic solid-phase extraction of polychlorinated biphenyls from environmental water samples. *Journal of Chromatography A*, **1304**, 241-245.

135. Zhang, W., Yan, Z., Gao, J., Tong, P., Liu, W., and Zhang, L. (2015) Metal–organic framework UiO-66 modified magnetite@silica core–

shell magnetic microspheres for magnetic solid-phase extraction of domoic acid from shellfish samples. *Journal of Chromatography A*, **1400**, 10-18.

136. Wang, X., and Deng, C. (2015) Preparation of magnetic graphene @polydopamine @Zr-MOF material for the extraction and analysis of bisphenols in water samples. *Talanta*, **144**, 1329-1335.

137. Su, H., Lin, Y., Wang, Z., Wong, Y.-L. E., Chen, X., and Chan, T.-W. D. (2016) Magnetic metal–organic framework–titanium dioxide nano-composite as adsorbent in the magnetic solid-phase extraction of fungicides from environmental water samples. *Journal of Chromatography A*, **1466**, 21-28.

138. Lin, S., Gan, N., Cao, Y., Chen, Y., and Jiang, Q. (2016) Selective dispersive solid phase extraction-chromatography tandem mass spectrometry based on aptamer-functionalized UiO-66-NH$_2$ for determination of polychlorinated biphenyls. *Journal of Chromatography A*, **1446**, 34-40.

139. Jia, Y., Su, H., Wong, Y.–L. E., Chen, X., and Chan, T.–W. D. (2016) Thermo-responsive polymer tethered metal-organic framework core-shell magnetic microspheres for magnetic solid-phase extraction of alkylphenols from environmental water samples. *Journal of Chromatography A*, **1456**, 42-48.

140. Du, F., Qin, Q., Deng, J., Ruan, G., Yang, X., Li, L., and Li, J. (2016) Magnetic metal–organic framework MIL-100(Fe) microspheres for the magnetic solid-phase extraction of trace polycyclic aromatic hydrocarbons from water samples. *Journal of Separation Science*, **39**(12), 2356-2364.

141. Yang, Y., Ma, X., Feng, F., Dang, X., Huang, J., and Chen, H. (2016) Magnetic solid-phase extraction of triclosan using core-shell Fe$_3$O$_4$@MIL-100 magnetic nanoparticles, and its determination by HPLC with UV detection. *Microchimica Acta*, **183**, 2467-2472.

142. Xie, Y., Deng, C., and Li, Y. (2017) Designed synthesis of ultra-hydrophilic sulfo-functionalized metal-organic frameworks with a magnetic core for highly efficient enrichment of the N-linked glycopeptides. *Journal of Chromatography A*, **1508**, 1-6.

143. Ma, X., Feng, F., Yang, Y., Dang, X., Huang, J., and Chen, H. (2017) Magnetic solid-phase extraction of N,N-diethyl-m-toluamide from baby toilet water prior to its HPLC–UV detection. *Journal of Chromatographic Science*, **55**(6), 662-668.

144. Liu, H., Chen, L., and Ding, J. (2017) A core-shell magnetic metal organic framework of type Fe$_3$O$_4$@ZIF-8 for the extraction of tetracycline antibiotics from water samples followed by ultra-HPLC-MS analysis. *Microchimica Acta*, **184**, 4091-4098.

145. Zhao, M., Xie, Y., Chen, H., and Deng, C. (2017) Efficient extraction of low-abundance peptides from digested proteins and simultaneous exclusion of large-sized proteins with novel hydrophilic magnetic

zeolitic imidazolate frameworks. *Talanta*, **167**, 392-397.

146. Ma, X., Zhou, X., Yu, A., Zhao, W., Zhang, W., Zhang, S., Wei, L., Cook, D. J., and Rou, A. (2018) Functionalized metal-organic framework nanocomposites for dispersive solid phase extraction and enantioselective capture of chiral drug intermediates. *Journal of Chromatography A*, **1537**, 1-9.

147. Xie, Y., Liu, Q., Li, Y., and Deng, C. (2018) Core-shell structured magnetic metal-organic framework composites for highly selective detection of N-glycopeptides based on boronic acid affinity chromatography. *Journal of Chromatography A*, **1540**, 87-93.

148. Deng, Y., Zhang, R., Li, D., Sun, P., Su, P., and Yang, Y. (2018) Preparation of iron-based MIL-101 functionalized polydopamine@Fe_3O_4 magnetic composites for extracting sulfonylurea herbicides from environmental water and vegetable samples. *Journal of Separation Science*, **41**(9), 2046-2055.

149. Khezeli, T., and Daneshfar, A. (2017) Development of dispersive micro-solid phase extraction based on micro and nano sorbents. *Trends in Analytical Chemistry*, **89**, 99-118.

6

Metal Organic Frameworks based Membranes for Pervaporation

Zhiqian Jia

Lab for Membrane Technology, College of Chemistry, Beijing Normal University, Beijing 100875, China

zhqjia@bnu.edu.cn

6.1 Introduction

Pervaporation (PV) is a promising technology for liquid separations. In PV, a liquid mixture contacts with the membrane, and the components are first absorbed by the membrane, followed by diffusion through the membrane and evaporation on the other side of the membrane, due to the driving force of vacuum or gas purge. Separation is achieved by the difference in the sorption and diffusion of the components in the membranes [1].

Pervaporation is usually employed for the removal of minor components in liquid mixtures, such as dehydration of organic solvents, removal of organic compounds from aqueous streams and separation of organic-organic mixtures. PV demonstrates incomparable advantages, especially in the separation of heat-sensitive, close-boiling and azeotropic mixtures due to the mild operating conditions, energy efficiency, ease of operation, eco-friendliness and high separation efficiency [2].

Metal organic frameworks (MOFs), composed of metal centers and organic linkers, have received remarkable attention recently due to their chemical versatility combined with designable framework topologies, high surface area and permanent porosity. MOFs have been used in gas storage [3], chemical separations [4], catalysis [5], adsorption [6,7], chromatography [8], drug delivery [9], pervaporation [10], etc. This chapter reviews the progress in MOFs-based PV membranes, including the screening criteria, continuous MOFs membranes and MOFs/polymer mixed matrix membranes (MMMs).

Metal Organic Frameworks, edited by Vikas Mittal
© 2019 Central West Publishing, Australia

6.2 Screening Criteria of MOFs

MOFs determine the PV performance of MOFs based membranes to a significant extent. In judicious selection of MOFs, the stability, hydrophobicity/hydrophilicity, pore size, dispersion and surface functional groups should be considered.

6.2.1 Stability

The stability of MOFs in liquids is crucial for PV. From the thermodynamic viewpoint, for liquids containing water, the free energy of the hydrolysis reaction of the metal ions determines the stability of MOFs [11]. For the reaction kinetics, the hydrolysis rate is related to the activation energy barrier. Generally, MOFs containing the group IV metals in the +4 oxidation state correlate with high stability. High metal coordination numbers create a crowding effect that prevents the formation of water clusters near the metal center [12]. Incorporation of the hydrophobic fluorinated and alkyl functional groups on the ligands can prevent the water molecules from adsorbing into the pores or clustering around the metal center, thus, improving the stability under humid conditions [13]. Ligands with high pKa values favor higher water tolerances. Metals and ligands with similar polarizability result in strong binding coordination [14]. In addition, the interpenetration of individual frameworks favors the generation of water-stable MOFs [15].

For example, MIL-100 is synthesized by octahedral M^{3+} trimers (M = Fe, Al, Cr) and 1,3,5-benzenetricarboxylate (BTC) ligand [16], and the highly charged trivalent metals lead to the strong metal-ligand bonds and excellent stability. MIL-101(Cr) is made up of chromium trimers and terephthalate linkers (BDC), and it can maintain BET surface area and PXRD pattern even after immersion in boiling water for 1 week [17]. UiO-66 possesses 12-coordinated zirconium-oxo clusters and is highly stable in water [18].

6.2.2 Hydrophobicity/Hydrophilicity

The hydrophobicity/hydrophilicity of MOFs is mainly determined by the ligands. For example, ZIF-8, ZIF-71, FMOF-1 (prepared by Ag^+ and 3,5-bis(trifluoromethyl)-1,2,4-triazolate) and MAF-2 (formed by Cu^{2+} and 3,5-diethyl-1,2,4-triazole) [19] are highly hydrophobic and exhibit substantial adsorption selectivity for organics over water [20].

Mg-MOF-74 (generated by Mg^{2+} and 2,5-dihydroxybenzene carboxylic acid) and ZIF-93 (synthesized by Zn^{2+} and 4-methyl-5imidazolecarboxaldehyde) are relatively hydrophilic. The hydrophobicity of MOFs can be characterized by the heat of water adsorption [21] or water uptake at low pressure [22].

6.2.3 Pore Size

The pore size of MOFs is an important parameter for PV. When the pore size is between the molecular kinetic diameters of two components, the molecules of the smaller component diffuse through the pores whereas the molecules of the larger component are excluded, thus, displaying molecular sieving effect. When the pore size is larger than the two components, the separation is achieved by the differences in the equilibrium adsorption [23].

MOFs are generally structurally flexible (e.g. breathing or gate opening), and the cut-off is usually not clear [24]. Thus, the high flexibility of MOFs hinders their application for shape-based molecular separations. In addition, the MOFs pores may be blocked by the polymer chains during MMMs fabrication, leading to the probable decrease in permeability and increase in selectivity. Confining a third agent (e.g. ionic liquid) in the MOFs pores can tune the pore size of MOFs accurately. The pore aperture can also be tuned by employing mixed linkers. By controlling the relative ratio of the linkers, a variety of pore sizes can be created [25].

6.2.4 Dispersion of MOFs

The dispersion of MOFs in polymer matrices determines the interfacial defects and PV performance. Several approaches for improving the dispersion of MOFs have been reported:

(i) Use of wet or pre-primed MOFs: newly prepared wet MOFs are directly dispersed in the polymer solutions to inhibit the aggregation of MOFs in the conventional drying process. Wet MOFs can also be pre-primed with a small amount of the polymer solution [26], thus, leading to the homogenous distribution of MOFs in polymers without agglomeration. It was reported that the newly prepared wet ZIF-8 obtained by centrifugation (without washing) can be well dispersed in water due to the positive surface potential, and there was no settlement even upon centrifugation at 8000 rpm [27].

(ii) Use of nanosized MOFs: large MOFs particles are prone to form sediments in casting polymer solutions. Thus, the use of nanosized MOFs is beneficial for achieving uniform dispersion.

(iii) Surface modification of MOFs: modification of MOFs with long-chain organic compounds may help to achieve good dispersion in the polymer solutions. Queen *et al.* [28] modified hydrophilic UiO-66-NH$_2$ with oleic acid via post-synthetic ligand exchange and observed good dispersion in hydrophobic cyclic olefin copolymer. Liu *et al.* [29] coated polydopamine on SO$_3$H-MIL-101(Cr) and added to poly(vinyl alcohol) (PVA) for dehydration of ethylene glycol. Compared with pure PVA membrane, the water permeability and selectivity of the composite membranes increased by 483% and 567% respectively due to the enhanced compatibility between the MOFs and PVA.

6.2.5 Surface Functional Groups

The surface functional groups of MOFs determine the interactions with the polymers and other components. The MOFs/polymer inter-action is related to the compatibility and interfacial defects. Using MOFs containing organic linkers similar to the polymer units or interacting strongly with the polymer can eliminate the interfacial defects and improve the compatibility and membrane selectivity. Hansen solubility parameters (HSP) of polymers and linkers can be used for selecting linkers for polymers [30,31].

To achieve the functionalization of MOFs, two methods have been reported [50]:

(1) use of functionalized ligands during preparation of MOFs. However, the functional groups may interfere with the structure of MOFs or may not be compatible with the synthetic conditions.

(2) post-synthetic modification (PSM) of MOFs. MOFs can be modified with different reagents after synthesis, thereby, generating topologically identical, but functionally diverse MOFs. Multiple functional units can also be introduced into a single framework by PSM.

6.3 Continuous MOFs Membranes

Continuous MOFs membranes are prepared by growing a thin layer of MOFs on porous substrates (ceramics, stainless steel, polymers, etc.). The substrates provide mechanical support and should exhibit minimal permeation resistance. Substrates containing the same metal as MOFs are favorable for reacting with the organic linkers and enhancing the adhesion between the MOF layers and substrates. For the synthesis of high-quality MOFs membranes, heterogeneous nucleation and growth of MOFs on the substrate should be controlled. Continuous MOFs membranes can be prepared by direct synthesis, secondary growth and counter diffusion methods.

In the direct synthesis, a porous substrate is immersed in a MOFs synthesis solution, and the MOFs layer grows directly on the substrate. Nevertheless, it is usually difficult to control the nucleation of MOFs, thus, resulting in the defective MOFs membranes with intercrystal voids.

Secondary growth, also called seeded growth, is commonly used in the preparation of MOFs membranes. In this process, MOFs crystals are firstly synthesized and deposited on the substrates as seeds using various methods (e.g. dip coating, rubbing of MOFs powder, etc.). Subsequently, secondary growth is conducted, leading to an effective control of the crystal growth and orientation of MOFs membranes.

In the counter-diffusion synthesis, two precursor solutions (e.g. a metal ion solution and an organic ligand solution) separated by a porous substrate are allowed to counter-diffuse through the substrate pores, and the nucleation and crystallization processes occur on the substrate [32].

Diestel *et al.* [33] observed that ZIF-8 membrane exhibited a separation factor of 8.4 for n-hexane/benzene. Lin and co-workers reported the pervaporation performance of ZIF-71 membranes and found separation factors of 6.07 and 21.38 for EtOH/water and MeOH/water, respectively [34]. Hu *et al.* [35] reported that the MIL-53 membrane exhibited a high separation factor (1317) for water/ethyl acetate, but only 1 and 1.2 for water/EtOH and water/t-BuOH, respectively. The authors prepared ZIF-71 membranes on α-Al_2O_3 hollow fibers and found a separation factor of 6.88 in a 5:95 ethanol/water mixture at 298 K, and the separation factor decreased to 5.02 at 328 K.

For continuous MOFs membranes, MOF pores should be the only permeation pathway through the membranes so that an excellent selectivity can be achieved in PV. Nevertheless, the membranes often contain intrinsic or extrinsic defects between crystalline domains, leading to nonselective flow [36]. Additionally, the MOFs layers are usually not robust enough to form membranes with large surface area.

6.4 MOFs/Polymer MMMs

Polymer membranes have been widely employed in PV, however, the membrane swelling usually leads to the deteriorated selectivity due to the plasticization effect. Therefore, MMMs, composed of the polymer and inorganic filler (e.g. zeolite, inorganic nanoparticles, etc.) phases, have been developed to enhance the permeation flux, selectivity and stability. However, the weak interfacial bonding and incompatibility between the inorganic fillers and polymeric matrices often result in the interfacial defects, declined selectivity and trade-off effect between the permeability and selectivity [37]. In comparison with inorganic fillers, MOFs are more compatible with the polymer matrices due to the presence of organic ligands. Thus, the addition of porous MOF fillers may improve the flux and selectivity, along with inhibiting the swelling of polymers. Numerous MOFs/polymer MMMs have been reported in literature in recent years for pervaporation purposes.

6.4.1 Solvent Dehydration

In solvent dehydration, the hydrophilic polymers and MOFs are often used. For examples, Jia *et al.* [38] modified MIL-53(Al)-NH$_2$ nanocrystals with formic acid, valeric anhydride and heptanoic anhydride. These were subsequently incorporated in poly(vinyl alcohol) for dehydration of ethanol. It was observed that the PV performance was tuned to anti-trade-off for formic acid and valeric anhydride modified MIL-53(Al)-NH$_2$. Similarly, NH$_2$-MIL-125(Ti)/NaAlg MMMs showed significant enhancement in the flux and selectivity during dehydration of acetic acid [39]. MIL-101(Cr)/polyimide Matrimid® MMMs displayed higher permeability and stability during dehydration of the reversible esterification reaction as compared to the pristine membrane [40].

UiO-66-NH$_2$/6FDA-HAB/DABA polyimide MMMs (loading of 0-20 wt%) showed improvements in the normalized flux and separation factor during isopropanol dehydration due to the effective compatibility between the UiO-66-NH$_2$ particles and 6FDA-HAB/DABA polyimide, along with the hydrophilicity of the amino groups of UiO-66-NH$_2$ [41]. EuBTB (BTB=benzene-1,3,5-tribenzoate), as a kind of the lanthanide-based two-dimensional MOFs (pores size of 0.5-0.8 nm), was utilized to prepare sodium alginate MMMs for the dehydration of 90 wt% ethanol. The membrane containing 5 wt% EuBTB exhibited an optimum performance (permeation flux of 1996 g/m^2 h and separation factor of 1160) and long-term stability at 350 K [42].

6.4.2 Organophilic Pervaporation

For the organophilic PV, the hydrophobic polymers and MOFs are usually employed. For instance, ZIF-8/silicone rubber membrane displayed high selectivity (10) and furfural permeability (104 Barrer) during the removal of furfural from aqueous solution at 80 °C [43]. In the recovery of bio-alcohols, ZIF-71/poly(dimethyl siloxane) (PDMS) membrane showed 100% increase in selectivity as compared to the pristine membrane, and the flux was also improved [44]. MIL-53/PDMS MMMs exhibited much higher flux (5467 g m^{-2} h^{-1}) than the pristine membrane (1667 g m^{-2} h^{-1}), while the separation factor remained at 11.1 during the ethanol recovery due to the ethanol-affinity channels of MIL-53 [45]. The PDMS membranes with 20% loading of ZIF-67 showed a 300% increase in the flux, along with a 200% enhancement in the separation factor, for ethanol/water mixtures as compared to unfilled PDMS membrane [46].

6.4.3 Organic-Organic Mixtures Separation

Organic-organic mixtures separation, e.g. benzene/cyclohexane, benzene/hexane, methyl tert-butyl ether (MTBE)/methanol(MeOH), etc., is important from the perspective of the chemical industry. The stability of the polymers and MOFs in organic mixtures should be considered first. For separating toluene/n-heptane mixture, Cu$_3$(BTC)$_2$/PVA membranes fabricated on the ceramic tubular substrate displayed higher separation factor (17.9) and permeate flux (133 g m^{-2} h^{-1}) than pristine PVA membrane (8.9, 14 g m^{-2} h^{-1}) due to the strong affinity of toluene to Cu$_3$(BTC)$_2$ [47]. In the separation of MeOH/MTBE mixture, [Cu$_2$(bdc)$_2$(bpy)]$_n$ was added to the sulfonated polyarylethersulfone with cardo (SPES-C), and the MMMs exhibited improved selectivity and flux (0.288 kg m^{-2} h^{-1}) owing to the preferential sorption and diffusion of MeOH on [Cu$_2$(bdc)$_2$(bpy)]$_n$ [48].

6.5 Conclusions

MOFs based pervaporation membranes exhibit significant potential for achieving outstanding performance due to the ease of design and modification of MOFs. In future, the (i) design and synthesis of stable and robust MOFs, (ii) tuning of the MOFs/polymer interactions, (iii) design of the microstructure of MMMs and (iv) optimization of the synergistic effects between the polymer matrices and MOFs should be studied further.

References

1. Jia, Z., and Wu, G. (2016) Metal-organic frameworks based mixed matrix membranes for pervaporation. *Microporous and Mesoporous Materials*, **235**, 151-159.
2. Shao, P., and Huang, R. Y. M. (2007) Polymeric membrane pervaporation. *Journal of Membrane Science*, **287**, 162-179.
3. Murray, L. J., Dincă, M., and Long, J. R. (2009) Hydrogen storage in metal-organic frameworks. *Chemical Society Reviews*, **38**, 1294-1314.
4. Li, J.-R., Kuppler, R. J., and Zhou, H.-C. (2009) Selective gas adsorption and separation in metal-organic frameworks. *Chemical Society Reviews*, **38**, 1477-1504.
5. Vilhelmsen, L. B., Walton, K. S., and Sholl, D. S. (2012) Structure and mobility of metal clusters in MOFs: Au, Pd, and AuPd clusters in MOF-74. *Journal of the American Chemical Society*, **134**, 12807-12816.
6. Jia, Z., Jiang, M., and Wu, G. (2016) Amino-MIL-53 (Al) sandwich-structure membranes for adsorption of p-nitrophenol from aqueous solutions. *Chemical Engineering Journal*, **307**, 283-290.
7. Paseta, L. Simón-Gaudó, E. Gracia-Gorría, F., and Coronas, J. (2016) Encapsulation of essential oils in porous silica and MOFs for trichloroisocyanuric acid tablets used for water treatment in swimming pools. *Chemical Engineering Journal*, **292**, 28-34.
8. Yusuf, K., Aqel, A., and Alothman, Z. (2014) Metal-organic frameworks in chromatography. *Journal of Chromatography A*, **1348**, 1-16.
9. Horcajada, P., Chalati, T., Serre, C., Gillet, B., Sebrie, C., Baati, T., Eubank, J. F., Heurtaux, D., Clayette, P., Kreuz, C., Chang, J.-S., Hwang, Y. K., Marsaud, V., Bories, P.-N., Cynober, L., Gil, S., Férey, G., Couvreur, P., and Gref, R. (2010) Porous metal-organic-framework nanoscale carriers as a potential platform for drug delivery and imaging. *Nature Materials*, **9**, 172-178.
10. Jue, M. L., Koh, D. Y., McCool, B. A., and Lively, R. P. (2017) Enabling widespread use of microporous materials for challenging organic solvent separations. *Chemistry of Materials*, **29**, 9863-9876.
11. Burtch, N. C., Jasuja, H., and Walton, K. S. (2014) Water stability and adsorption in metal-organic frameworks. *Chemical Reviews*, **114**,

10575-10612.

12. DeCoste, J. B., Peterson, G. W., Schindler, B. J., Killops, K. L., Browe, M. A., and Mahle, J. J. (2013) The effect of water adsorption on the structure of the carboxylate containing metal-organic frameworks Cu-BTC, Mg-MOF-74, and UiO-66. *Journal of Materials Chemistry,* **A1**, 11922-11932.

13. Taylor, J. M. Vaidhyanathan, R. Iremonger, S. S., and Shimizu, G. K. H. (2012) Enhancing water stability of metal-organic frameworks via phosphonate monoester linkers. *Journal of the American Chemical Society,* **134**, 14338-14340.

14. Pearson, R. G. (1963) Hard and soft acids and bases. *Journal of the American Chemical Society,* **85**, 3533-3539.

15. Jasuja, H., and Walton, K. S. (2013) Effect of catenation and basicity of pillared ligands on the water stability of MOFs. *Dalton Transactions,* **42**, 15421-15426.

16. Cunha, D., Ben Yahia, M., Hall, S., Miller, S.R., Chevreau, H., Elkaïm, E., Maurin, G., Horcajada, P., and Serre, C. (2013) Rationale of drug encapsulation and release from biocompatible porous metal-organic frameworks. *Chemistry of Materials,* **25**, 2767-2776.

17. Seo, Y.-K., Yoon, J. W., Lee, J. S., Hwang, Y. K., Jun, C.-H., Chang, J.-S., Wuttke, S., Bazin, P., Vimont, A., Daturi, M., Bourrelly, S., Llewellyn, P. L., Horcajada, P., Serre, C., and Férey, G. (2012) Energy-efficient dehumidification over hierachically porous metal-organic frameworks as advanced water adsorbents. *Advanced Materials,* **24**, 806-810.

18. Cavka, J. H., Jakobsen, S., Olsbye, U., Guillou, N., Lamberti, C., Bordiga, S., and Lillerud, K. P. (2008) A new zirconium inorganic building brick forming metal organic frameworks with exceptional stability. *Journal of the American Chemical Society,* **130**, 13850-13851.

19. Zhang, J. P., and Chen, X.-M. (2008) Exceptional framework flexibility and sorption behavior of a multifunctional porous cuprous triazolate framework. *Journal of the American Chemical Society,* **130**, 6010-6017.

20. Lee, J. Y., Olson, D. H., Pan, L., Emge, T. J., and Li, J. (2007) Microporous metal-organic frameworks with high gas sorption and separation capacity. *Advanced Functional Materials,* **17**, 1255-1262.

21. Küsgens, P. Rose, M. Senkovska, I. Fröde, H. Henschel, A. Siegle, S., and Kaskel, S. (2009) Characterization of metal-organic frameworks by water adsorption. *Microporous and Mesoporous Materials,* **120**, 325-330.

22. Ng, E.-P., and Mintova, S. (2008) Nanoporous materials with enhanced hydrophilicity and high water sorption capacity. *Microporous and Mesoporous Materials,* **114**, 1-26.

23. Wee, L. H., Li, Y., Zhang, K., Davit, P., Bordiga, S., Jiang, J., Vankelecom, I. F. J., and Martens, J. A. (2015) Submicrometer-sized ZIF-71 filled

organophilic membranes for improved bioethanol recovery: mechanistic insights by Monte Carlo simulation and FTIR Spectroscopy. *Advanced Functional Materials*, **25**, 516-525.

24. Férey, G., and Serre, C. (2009) Large breathing effects in three-dimensional porous hybrid matter: facts, analyses, rules and consequences. *Chemical Society Reviews*, **38**, 1380-1399.

25. Thompson, J. A., Blad, C. R., Brunelli, N. A., Lydon, M. E., Lively, R. P., Jones, C. W., and Nair, S. (2012) Hybrid zeolitic imidazolate frameworks: controlling framework porosity and functionality by mixed-linker synthesis. *Chemistry of Materials*, **24**, 1930-1936.

26. Hua, D., Ong, Y. K., Wang, Y., Yang, T., and Chung, T.-S. (2014) ZIF-90/P84 mixed matrix membranes for pervaporation dehydration of isopropanol. *Journal of Membrane Science*, **453**, 155-167.

27. Jia, Z., Wu, G., Wu, D., Tong, Z., and Ho, W. S. (2017) Preparation of ultra-stable ZIF-8 dispersions in water and ethanol. *Journal of Porous Materials*, **24**, 1655-1660

28. Bae, Y. J., Cho, E. S., Qiu, F., Sun, D. T., Williams, T. E., Urban, J. J., and Queen, W. L. (2016) Transparent metal-organic framework/polymer mixed matrix membranes as water vapor barriers. *ACS Applied Materials and Interfaces*, **8**, 10098-10103.

29. Zhang, W., Ying, Y., Ma, J., Guo, X., Huang, H., Liu, D., and Zhong, C. (2017) Mixed matrix membranes incorporated with polydopamine-coated metal-organic framework for dehydration of ethylene glycol by pervaporation. *Journal of Membrane Science*, **527**, 8-17.

30. Hansen, C. M. (2004) 50 Years with solubility parameters-past and future. *Progress in Organic Coatings*, **51**, 77-84.

31. Seoane, B., Coronas, J., Gascon, I., Benavides, M. E., Karvan, O., Caro, J., Kapteijn, F., and Gascon, J. (2015) Metal-organic framework based mixed matrix membranes: a solution for highly efficient CO_2 capture? *Chemical Society Reviews*, **44**, 2421-2454.

32. Yao, J. F., Dong, D. H., Li, D., He, L., Xu, G. S., and Wang, H. T. (2011) Contra-diffusion synthesis of ZIF-8 films on a polymer substrate. *Chemical Communications*, **47**, 2559-2561.

33. Diestel, L., Bux, H., Wachsmuth, D., and Caro, J. (2012) Pervaporation studies of n-hexane, benzene, mesitylene and their mixtures on zeolitic imidazolate framework-8 membranes. *Microporous and Mesoporous Materials*, **164**, 288-293.

34. Dong, X., and Lin, Y. S. (2013) Synthesis of an organophilic ZIF-71 membrane for pervaporation solvent separation. *Chemical Communications*, **49**, 1196-1198.

35. Hu, Y., Dong, X., Nan, J., Jin, W., Ren, X., Xu, N., and Lee, Y. M. (2011) Metal-organic framework membranes fabricated via reactive seeding. *Chemical Communications*, **47**, 737-739.

36. Choi, J., Jeong, H. K., Snyder, M. A., Stoeger, J. A., Masel, R. I., and Tsapatsis, M. (2009) Grain boundary defect elimination in a zeolite

membrane by rapid thermal processing. *Science*, **325**, 590-593.

37. Kariduraganavar, M. Y., Varghese, J. G., Choudhari, S. K., and Olley, R. H. (2009) Organic– inorganic hybrid membranes: solving the trade-off phenomenon between permeation flux and selectivity in pervaporation. *Industrial and Engineering Chemistry Research*, **48**, 4002-4013.

38. Wu, G., Jiang, M., Zhang, T., and Jia, Z. (2016) Tunable pervaporation performance of modified MIL-53 (Al)-NH2/poly (vinyl alcohol) mixed matrix membranes. *Journal of Membrane Science*, **507**, 72-80.

39. Su, Z., Chen, J. H., Sun, X., Huang, Y., and Dong, X. (2015) Amine-functionalized metal organic framework (NH2-MIL-125(Ti)) incorporated sodium alginate mixed matrix membranes for dehydration of acetic acid by pervaporation. *RSC Advances*, **5**, 99008-99017.

40. De la Iglesia, Ó., Sorribas, S., Almendro, E., Zornoza, B., Téllez, C., and Coronas, J. (2016) Metal-organic framework MIL-101 (Cr) based mixed matrix membranes for esterification of ethanol and acetic acid in a membrane reactor. *Renewable Energy*, **88**, 12-19.

41. Xu, Y. M., Japip, S., and Chung, T.-S. (2018) Mixed matrix membranes with nano-sized functional UiO-66-type MOFs embedded in 6FDA-HAB/DABA polyimide for dehydration of C1-C3 alcohols via pervaporation. *Journal of Membrane Science*, **549**, 217-226.

42. Gao, B., Jiang, Z., Zhao, M., Wu, H., Pan, F., Mayta, J. Q., Chang, Z., and Bu, X. (2018) Enhanced dehydration performance of hybrid membranes by incorporating lanthanide-based MOFs. *Journal of Membrane Science*, **546**, 31-40.

43. Liu, X., Jin, H., Li, Y., Bux, H., Hu, Z., Ban, Y., and Yang, W. (2013) Metal-organic framework ZIF-8 nanocomposite membrane for efficient recovery of furfural via pervaporation and vapor permeation. *Journal of Membrane Science*, **428**, 498-506.

44. Li, Y., Wee, L. H., Martens, J. A., and Vankelecom, I. F. J. (2014) ZIF-71 as a potential filler to prepare pervaporation membranes for bio-alcohol recovery. *Journal of Materials Chemistry A*, **2**, 10034-10040.

45. Zhang, G., Li, J., Wang, N., Fan, H., Zhang, R., Zhang, G., and Ji, S. (2015) Enhanced flux of polydimethylsiloxane membrane for ethanol permselective pervaporation via incorporation of MIL-53 particles. *Journal of Membrane Science*, **492**, 322-330.

46. Khan, A., Ali, M., Ilyas, A., Naik, P., Vankelecom, I. F.J., Gilani, M. A., Bilad, M. R., Sajjad, Z., and Khan, A. L. (2018) ZIF-67 filled PDMS mixed matrix membranes for recovery of ethanol via pervaporation. *Separation and Purification Technology*, **206**, 50-58.

47. Zhang, Y., Wang, N., Ji, S., Zhang, R., Zhao, C., and Li, J.-R. (2015) Metal-organic framework/poly (vinyl alcohol) nanohybrid membrane for the pervaporation of toluene/n-heptane mixtures. *Journal of Membrane Science*, **489**, 144-152.

48. Han, G. L., Zhou, K., Lai, A. N., Zhang, Q. G., Zhu, A. M., and Liu, Q. L.

(2014) [Cu$_2$(bdc)$_2$(bpy)]$_n$/SPES-C mixed matrix membranes for separation of methanol/methyl tert-butyl ether mixtures. *Journal of Membrane Science*, **454**, 36-43.

7

Development of Computational Tools for Diverse Applications of Metal Organic Frameworks: Challenges and Outlooks

Bukhvalov Danil,[a] Pawan Kumar,[b,*] Abhinav Gondhi,[b] Tapta Kanchan Roy[c] and Ki-Hyun Kim[d,*]

[a]Department of Chemistry, Hanyang University, 222 Wangsimni-Ro, Seoul 04763, Republic of Korea
[b]Department of Nano Science and Materials, Central University of Jammu, 181143, Jammu, India
[c]Department of Chemistry and Chemical Science, Central University of Jammu, 181143, Jammu, India
[d]Department of Civil & Environmental Engineering, Hanyang University, 222 Wangsimni-Ro, Seoul 04763, Korea

*Corresponding authors: kkim61@hanyang.ac.kr; pawannano10@gmail.com

7.1 Introduction

Computational tools have been widely employed over the past few decades to examine the structure and properties of materials. The first theoretical calculation in materials chemistry was derived by W. Heilter and F. London in 1927 [1]. Subsequently, a solution to the wave equation for complex atomic systems and the first semi-empirical atomics orbital calculations were derived in the 1940s and 1950s, respectively [1]. Since then, the first polyatomic calculation using Gaussian orbitals (1950), the first *ab initio* Hartree-Fock methods (1956), calculation of electron energies using molecular orbital *Huckel* methods (1964), molecular mechanics - (*MM2*) force fields (1970) and the first molecular orbital calculations (1980) have tremendously advanced materials science and technology [1,2]. Molecular mechanics, semi-empirical mechanics and quantum mechanics are the three fundamental categories of computational chemistry utilized in materials science and technology.

Metal Organic Frameworks, edited by Vikas Mittal
© 2019 Central West Publishing, Australia

Density functional theory (DFT)-based simulation/computational tools are commonly used to determine the design, structure and properties of materials. DFT was developed from quantum mechanics on the basis of the time-independent Schrodinger equation to find minimum energy structures, binding energies and details of electronic structures [3]. DFT is an established method that facilitates the design of novel materials without the need for cost-intensive, time-consuming or impractical experiments. More specifically, simulation/computational tools can predict properties (i.e., chemical structure, binding energies, adsorption isotherms, diffusion coefficients, optical, electrical, mechanical properties, and many more) of advanced materials such as metal organic frameworks (MOFs), graphene and hybrid materials [4,5].

In this chapter, we explore the computational tools used in diverse applications of MOFs and describe the present challenges and future outlooks. MOFs are a new class of porous materials with extraordinary properties such as large internal surface area, high porosity and high degree of crystallinity; these materials can be designed using flexible rational design through control of the architecture and functionalization of pores [6]. In light of these merits, several devices incorporating MOFs have been employed to detect small molecules, solvents pesticides, explosives, biological markers, etc. [7,8]. It is, however, not yet feasible to synthesize and test a large number of MOFs for every application of interest [9]. There is, hence, a strong demand for computational modeling to assess the capabilities of MOFs in various respects. High-throughput computational screening methods are, thus, the need of the hour for identification of promising candidates for different applications. Such efforts will ultimately help develop useful structural-property relations through pre-evaluation of the properties of thousands of MOFs [3]. Computational studies can potentially offer deep insights into phenomena occurring at the microscopic level [9]. Several computational tools are now used preferably for experimental characterization of MOFs.

7.2 Experimental Aspects of MOFs and Diverse Applications

The general approach used in computational chemistry is to reproduce the values obtained through experiments. This step requires the construction of models and choice of contemporary computational chemistry tools. The second step involves using these models and methods to predict future experimental results and propose novel materials. We, therefore, provide a brief description of experimental results obtained for MOFs.

MOFs have been largely developed for gas separation, gas storage and sensing of small molecules and explosives [8]. MOFs are useful to detect a change in luminescence due to a change in porosity and host-guest interactions, which makes them excellent candidates for sensing applications. Three different categories of MOF sensors currently exist: luminescence-based sensors, electrochemical-based sensors and electromechanical-based sensors. For the detection of specific molecules, some functionalities should be endowed. Three different strategies have been proposed and employed to modify MOFs: modification with specific organic ligands or doping of the frame with metal ions [6], post-synthesis modification (PSM) [10] and capture of different functional molecules and nanoparticles (NPs) within the framework.

The nanoscale luminescent MOF1 [(Eu$_2$) (BDC)$_3$. (H$_2$O)$_2$.(H$_2$O)$_2$] was devised to sense nitroaromatic explosives in ethanol solution by determining the photoluminescence properties of the material [11]. Nitroaromatic compounds had a considerable quenching effect on the luminescence intensity of MOF1 in ethanol solution, indicating that nanoscale MOF1 is a suitable candidate for sensing a small amount of nitroaromatic explosives [12]. Wang *et al.* [13] developed amino-functionalized metal organic frameworks (NH$_2$–Cu$_3$(BTC)$_2$; BTC=benzene-1,3,5-tricarboxylate) as electrochemical sensors of lead. An anodic voltammetric technique was used to detect lead. Voltammograms (graph of peak current vs. lead concentration) were recorded using an Ag/AgCl electrode under the following conditions: 0.1 mol of acetate buffer solution (pH= 4.5), differential pulse stripping voltammetric potential from -1.0 to -0.1 V, step potential of 10 mV, pulse amplitude of 100mV and pulse width of 1.0 s. Kumar *et al.* [14] used Basolite-1200 with protonated doping to sense Mecoprop by applying a potential to the thin film substrate. The electrical conductivity of the PMOF substrate in response to increasing amounts of Mecoprop was measured. The conductivity of the PMOF substrate increased with an increase in the amount of the Mecoprop, indicating that PMOF substrate is useful for conductometric sensing of Mecoprop. Chen *et al.* [7] reported a luminescent MOF, Tb (BTC) ·G (MOF-76, G = guest solvent), for sensing of anions (fluorine (F$^-$), chlorine (Cl$^-$), and bromine (Br$^-$)). The luminescence properties of anions with Tb (BTC)·G (MOF-76b, G = methanol) were studied in methanol solutions containing different concentrations of NaX (X = F$^-$, Cl$^-$ and

Br$^-$) and Na$_2$X (X = CO$_3$$^{2-}$ and SO$_4$$^{2-}$). When MOF-76b was exposed to F$^-$, there was a considerable increase in the intensity of luminescence, suggesting that MOF-76b can be used as a high sensitivity sensor for F$^-$.

Based on these experimental results, there are three main areas of interest in theoretical modeling [7]. The first area involves evaluating changes in the optical properties of MOF as a result of adsorption of molecules; the second involves assessing the influence of guest molecules on the chemical properties of MOFs, while the third area of interest is assessing changes in the porosity of MOFs caused by adsorption of molecules.

7.3 DFT-Based Calculations

DFT is a first-principle based on quantum chemistry method. It is commonly used to simulate the atomic and electronic structures of various compounds with their properties. DFT-based methods can be used for large systems with good accuracy and is suitable for modeling various chemical processes including those important for the evaluation of sensing properties. Hence, we briefly introduce the general aspects of DFT-based methods and some related issues concerning modeling of the interactions of MOFs with guest molecules.

7.3.1 Basic Principles of DFT Calculations

In the absence of time-dependent external fields and within the framework of Born-Oppenheimer approximation, a system for which only the atomic composition and positions are known can be described by the many-body time-independent Schrödinger equation: $\hat{H}\mathbf{\Psi} = E\mathbf{\Psi}$. In this equation, E is the total energy of the system, $\mathbf{\Psi}$ is the wave function, and the caret denotes a differential operator (mathematical function such as a derivative, Laplasian, etc.). The Hamiltonian, $\hat{H}$, includes the kinetic energies of the nuclei, $\hat{T}_n$ and the electrons, $\hat{T}_e$, as well as the Coulomb interactions between nuclei and electrons (V_{ne}), among electrons (V_{ee}) and among nuclei (V_{nn}):

$$\hat{H} = \hat{T}_e + \hat{T}_n + V_{ne} + V_{ee} + V_{nn}$$

The wave function, $\mathbf{\Psi}$, contains all pieces of information that can be extracted from the system within limits set by the Heisenberg uncertainty principle. However, this equation is too complicated to be solved exactly, except for very small systems (such as molecules comprising only a few atoms). For this reason, most modern computational approaches adopt some simplifying approximations.

The first step toward modeling of realistic systems is the Born-Oppenheimer approximation. This approach is valid when all excited electronic states are much higher in energy than the ground state. This structure of energy levels corresponds to stable atomic structures. Based on this approximation, nuclei can be considered unmovable. By excluding nuclei-nuclei interactions from the Schrödinger equation, only the motion of electrons in the field created by the nuclei and other electrons can be considered [15]. The next approach is Hartree-Fock approximation, which can help reduce the many-body problem to the one-electron problem by considering the motion of each given electron in the field created by all nuclei and other electrons [16]. This field is numerically described by its potential. Note that this potential is the result of several approximations.

DFT methods are based on two theorems proposed by Hohenberg and Kohn (1964). The first of the Hohenberg–Kohn theorems shows that there is a unique correspondence between electron density, $\rho(\mathbf{r})$ and the potential in which the electrons move [17]. The density, therefore, uniquely determines the ground-state wave function and vice versa. Thus, for a system, there is a universal functional of the density that defines its total energy $(E[\rho_0(\mathbf{r})] = E_0)$. According to the second Hohenberg–Kohn theorem, the energy corresponding to the ground state level can be obtained by variational methods: the density that minimizes the total energy is the exact ground state density [17].

7.3.2 Scheme of DFT-Based Calculations

Running simulations using DFT-based codes requires description of the atomic structure (full list of atoms and their coordinates $(\mathbf{r})$) and the exchange-correlation functional $(E_{xc}[\rho(\mathbf{r})])$, which is an estimation of electron-electron interactions. Additionally, required potentials or pseudopotentials (see below) are specially generated for the selected kind of exchange-correlation functional. These can be generated with additional codes distributed with packages or uploaded from web libraries. Note that different programs use different types and standards of (pseudo)potentials.

As the first step of running DFT-based code, the distribution of electron densities is calculated using initial values of $\rho_{in}(r)$, determined for single atoms of each type and atomic positions. In the second step, the previously calculated electron densities are used to calculate electro-static and exchange-correlation terms of the Hamiltonian. The third step is numerical solution of the Schrödinger equation. This step is the most computationally intensive. Last, new electron density ($\rho_{out}(\mathbf{r})$) values are calculated and compared with the initial values. If the differences are smaller than the adjusted values, electron densities corresponding to the ground state are obtained. If the differences between the initial and final electron density values are larger than the adjusted value, $\rho_{out}(\mathbf{r})$ is used as the new initial value for the next iteration. To stabilize the numerical calculations, a mixing scheme described by the equation below can be used:

$$\rho_{in}^{(\mu+1)} = \alpha\rho_{out}^{(\mu)} + (1 - \alpha)\rho_{in}^{(\mu)}$$

where $\rho_{in/out}^{(\mu)}$ is the electronic density at input or output of the μth iteration, and α is the mixing coefficient. The use of small values for the mixing coefficient not only increases the stability of numerical calculations, but also the time it takes to obtain the ground state electron density. In the case of porous systems and surfaces, stabilization of numerical calculations by decreasing the mixing coefficient is of practical importance.

There are two different types of approaches within DFT. One is representation of the system as a molecule or several molecules in a void. In contrast, the other models the system as periodic with a Bloch wave function:

$$e^{ikr}u(\mathbf{r})$$

where u is a periodic function with the same periodicity as the crystal, and $\mathbf{k}$ is a vector of real numbers called a crystal wave vector.

Both approaches have been used to model the interactions between MOFs and molecules. In the first approach, only part of the MOF (usually the metal-oxide core with ligand) is considered (Figure 7.1) [19]. This permits fast calculations, especially when guest molecules interact with open metal sites [13,18-19]. However, this approach cannot evaluate the effects of pore size and shape. The interaction between the guest molecule and ligands are also not determinable using this approach. These problems can be solved by building large molecule-like structures that contain several nodes of the MOF.

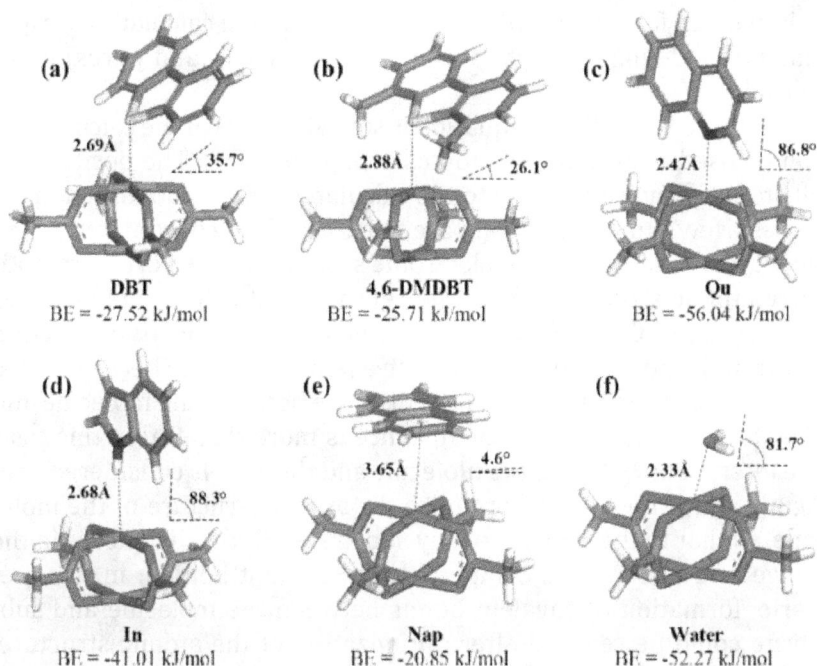

(a)

2.69Å   35.7°

DBT
BE = -27.52 kJ/mol

(b)

2.88Å   26.1°

4,6-DMDBT
BE = -25.71 kJ/mol

(c)

2.47Å   86.8°

Qu
BE = -56.04 kJ/mol

(d)

2.68Å   88.3°

In
BE = -41.01 kJ/mol

(e)

3.65Å   4.6°

Nap
BE = -20.85 kJ/mol

(f)

2.33Å   81.7°

Water
BE = -52.27 kJ/mol

Figure 7.1 Adsorption configuration and binding energies of various compounds at the Cu site of Cu-BTC. Reproduced from Reference 19b with permission from American Chemical Society.

There are many periodic codes that can be used in DFT. Full-potential codes (such as Wien 2k) are usually applied to model bulk solids [20,21]. Calculation of adsorption on the surface and molecule-like structures using full-potential methods requires prolonged computational time and memory. Modeling of the interaction between molecules and surfaces relies on methods that represent wave functions using so-called pseudopotentials [22]. Pseudopotentials can replace the complicated effects of the motion of non-valence electrons of an atom and its nucleus (core) with effective potentials or pseudopotentials. Thus, the Schrödinger equation contains a model effective potential term instead of a Coulombic potential term for core electrons. In other words, the number of electrons considered can be decreased. For example, in the case of a silicon atom, instead of performing calculations for 14 electrons (*1s2 2s2 2p6 3s2 3p2*), only four valence electrons (*3s2 3p2*) are considered in the calculations. The most commonly used pseudo-potential based codes are VASP, [23,24], ABINIT [25], SIESTA [26] and CASTEP [27]. These codes result in much faster calculations than full-potential calculations, especially if modeling of empty space (surfaces, voids, and pores) is required.

Once the Schrödinger equation is solved, results of the calculations can be used to evaluate the forces acting on atoms. The positions of all atoms change according to the calculated values and directions of forces. New atomic positions are used as initial coordinates in the new cycle of calculation of electronic structure. These cycles to optimize atomic structure should be repeated until the value becomes smaller than 0.05 eV/Å. In the case of molecular adsorption, selection of the initial position of a molecule in a pore is crucial. There are three possible scenarios in terms of the distance between the molecule and the atoms of the MOF: (i) the distance is more than 5Å; in this case, the interaction between the molecule and the MOF is considered negligible and is not used to optimize the atomic structure of the molecule, similar to the absence of any substrate; (ii) the distance is in the range of 1~1.5 Å (e.g., comparable to covalent bonds); in this scenario, formation of covalent bonds between the molecule and substrate could be realized after optimization of the atomic structure; (iii) the distance is ~ 2 Å; in this case, further optimization of atomic positions could yield the structure corresponding to non-covalent adsorption [10]. Building the proper set of initial coordinates is, thus, very important to construct a model that is consistent with the experimental results.

Optimized atomic structures, electrons densities and the Hamiltonian obtained from the calculations can be used for further calculation of

- total energy, which can be used to calculate the enthalpy of adsorption;
- densities of states (electronic structure);
- values of the magnetic moments on atoms.

Other characteristics of the system that can be calculated from the Hamiltonian and atomic positions include:

- phonon frequencies and eigenvectors at any wave vector;

- full phonon dispersions;
- effective charges and dielectric tensors;
- electron-phonon interactions;
- infrared and (non-resonant) Raman cross-sections;
- EPR and NMR chemical shifts;
- X-ray absorption spectra.

At the current stage of hardware development, calculations for a system with 200-300 atoms can be performed within a reasonable time (1~10 days) on local computer clusters without using super-computers. The exact calculation time depends on the stability of numerical calculations and the number of orbitals; for instance, heavier elements with *d*- and *f*-orbitals might require longer computational times than lighter elements.

7.3.3 Determination of the Electronic and Optical Properties of MOFs

It is important to evaluate how DFT-based tools can be used to simulate the above-described processes when investigating MOFs as chemical and/or biological sensors. First, the change in luminescent properties of adsorbed molecules should be considered. As DFT relies on an approximation for the description of electron-electron interactions, band gap values (or HOMO/LUMO gap in molecular systems) [29] and the energy of London dispersion forces (so-called van der Waals interactions) are underestimated [30].

To determine optical properties, the value of the band-gap is crucial, as it controls the wavelength of adsorbed or emitted light. There are two different approaches to obtain the correct value of the bandgap from DFT-based programs. The first one involves the use of so-called hybrid functionals instead of standard ones. Hybrid functionals are approximations to the exchange-correlation energy functional in DFT. Hybrid functionals contain a component of exact exchange from Hartree–Fock theory with exchange and correlation based on non-hybrid approaches such as LDA (*Local-density approximations*) or GGA (*generalized gradient approximation*) [31]. The most regularly used types of hybrid functionals are now implemented in the latest versions of all propagated DFT based codes. Unfortunately, there are no strict rules for selection of the type of hybrid functional. This choice can be made only when the exact atomic structure of a studied compound without impurities or adsorbed molecules is known. Another approach is the GW approach, which provides the first principle calculations for the value of the band gap by considering many-body effects. This method uses wavefunctions and self-energies of the system, calculated by DFT to perform many-body calculations based on a Green's function (G) method with screened Coulomb interactions (W) [29]. Unfortunately, in the case of MOFs, the number of the atoms per unit cell is very high (> 100), while values can be calculated in a reasonable time only for cells with 20-30 atoms. DFT-based methods can provide accurate information about changes in the values of the band gap and the magnitude of such changes [32]. In a recent experimental work, changes in the electronic structure of Fe-MOF-74 were observed after adsorption of molecular oxygen [33]. Another study demonstrated that p-type doping of $Co_3(NDC)_3MOF$ resulted in adsorption of iodine molecules on the linkers [34]. Changes in the band gap of UiO-66 MOF and adsorbed molecules were described theoretically and detected experimentally by measuring luminescence quenching [35].

7.3.4 Energetics of Adsorption

Another disadvantage of DFT-based methods is the failure to correct for weak interactions. London dispersion forces originate from the electric field created by reciprocal polarization of two non-polar molecules. The ability to polarize is related to electron-electron interactions [36]. Some functionals used in DFT (such as local density approximation (LDA)) can provide reasonably correct estimates of the energies and distances related to weak interactions [36]. However, this correctness is associated with so-called 'error cancellation'. More accurate types of functionals (such as generalized gradient approximation (GGA)) and its further derivative scan provide a better description of covalent bonds, but do not consider London dispersion forces. In recent decades, several corrections have been introduced into GGA functionals to allow more accurate evaluation of weak forces [30,37]. However, there is a lack of experimental measurements of weak forces that can be used to fit computational schemes.

The potential of DFT-based programs with the described corrections to model the adsorption of small molecules in MOFs has already been discussed in several reviews [5,38]. Fortunately, when modeling sensing properties, factors related to London dispersion forces are not very significant. Significant changes in the physical and chemical properties of both the MOF and guest molecules can lead to the formation of robust chemical bonds. Molecule-MOF interactions with charge transfer can lead to the formation of donor-acceptor bonds. The interaction of the guest molecule with an open metal site [39] then facilitates the formation of new coordination bonds. Experimental and theoretical studies [18,40-42] have provided evidence that the binding energy (enthalpy of adsorption) between large molecules and MOFs is on the order of 30-60 kJ/mol; this is an order of magnitude higher than the energy of London dispersion forces (~1-5 kJ/mol). Therefore, some inexactness in DFT calculations of so-called "van der Waals corrections" has a negligible effect when modeling the sensing properties of MOFs.

The next property of MOFs influenced by adsorption of molecules is chemical activity, which is also essential for sensing applications. DFT can model different redox reactions fairly accurately. All DFT-based programs listed above have been used to calculate oxygen reduction reactions in both acid and alkaline environments [43,44]. Porphyrins have been employed as catalysts in scenarios where open metal sites also served as the source of chemical activity [45]. Other types of catalytic activities of surfaces and molecules have also been modeled successfully using DFT. Despite the current lack of theoretical modeling of catalytic properties of MOFs [46], many models and methods have been developed to assess the influence of guest molecules on the chemical activity of porous materials.

7.4 Beyond DFT

7.4.1 Molecular Dynamics

Changes in the optical, electrical and mechanical features of MOFs also need to be considered. Robust adsorption of relatively large guest molecules on open metal sites decreases the volume of the void inside the MOF (Figure 7.2) [16]. DFT-based methods are useful to evaluate the geometric structure of the adsorbed molecule and to assess the decrease in pore size with increasing concentration of guest. However, DFT approach does not consider the motion of molecules at ambient conditions (computation is performed at zero Kelvin temperature). As the binding energy between molecules and MOFs in the discussed case is rather large, the chemical bonds will be stable at room temperature [47]. However, modeling of the exact position of the guest molecule should consider temperature. Molecular dynamic simulations could provide a better description of the effect of guest molecules on the porosity of MOFs. There are two branches of molecular dynamics (MD): first-principles MD and model-based MD.

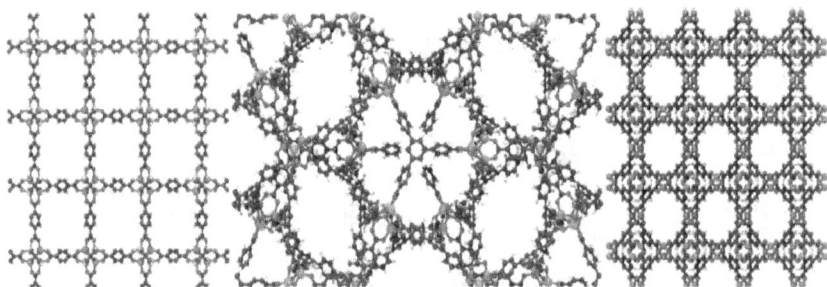

Figure 7.2 Three common metal organic frameworks investigated for their gas storage properties: IRMOF-1 (left), MOF-177 (middle), and HKUST-1 (right). Reproduced from Reference 16 with permission from American Chemical Society.

The first principle of MD is a computer simulation approach used to study the physical actions of condensed phase and molecules. In this method, atoms and molecules interact for a fixed interval of time (usually femtoseconds) for dynamic development of the modeled system. Molecular mechanics force fields are used to calculate the forces between the atoms and their potential energy, while taking into account the contribution of temperature. These methods are now routinely used to investigate the structure, dynamics and thermodynamics of biological molecules and their complexes [47]. They are also used in the determination of structures from X-ray crystallography and NMR experiments. The adsorption of CO, N_2, and CO_2 to Mg-MOF-74 was calculated using *ab initio* periodic DFT-D calculations [30] with the hybrid B3LYP functional [48-50]. All calculations were performed using CRYSTAL code first-principle MD packages. A limitation of this approach is the time required for calculations. For correct and numerically stable calculations, very small time steps (usually 1~10 fs) should be used. Therefore, for realistic timescales (e.g., milliseconds), hundreds of thousands of cycles of calculations are required. At the current level of hardware development, first-principles MD is usually employed only to model annealing processes.

A non-first-principles approach to MD was first introduced by Alder and Wainwright in 1957, to study the interactions of hard spheres [18]. MD methods can generate a series of time-correlated points in phase space (a trajectory) by propagating a starting set of coordinates and velocities according to Newton's second equation in a series of finite time steps [18]. Greathouse *et al.* [19] reported the interaction of water with MOF-5 using the MD method. The authors showed that the lower was the value of water, the more stable was the MOF-5 framework. When the content of water increased to more than 4%, the MOF became unstable. This was because of the sensitivity of the zinc-based MOFs to water; the MOFs lost their high surface area when they came in contact with water. The closeness of the energy interaction values between Zn-ions and water as well as Zn-ions and oxygen atoms from ligands explained why water molecules were able to enter pores easily and disturb the framework [53]. MD approaches are usually employed to calculate isotherms of adsorption of various small molecules (H_2, CO_2, CH_4) to MOFs [12].

A strength of MD-based methods is the ability to model systems comprising hundreds or thousands of atoms for an extended period of time (seconds). For adsorption of molecules on MOFs, MD methods permit variation in concentration of adsorbed molecules. The role of water and other solvents can also be considered. However, intra molecular hydrogen bonds are not properly described in modern force fields [53]. Furthermore, van der Waals interactions in MD are generally described by Lennard-jones Potentials, which are based on the Fritz London theory that is only valid in vacuum. This could be an issue if there is a high concentration of adsorbed molecules.

In the context of evaluating the sensing properties of MOFs, MD-based methods can be used to evaluate the influence of adsorption of molecules to pores at different temperatures [9,50-54]. Another application of these methods is using the coordinates from MD-simulations as input for DFT-based calculations of the bandgap and modeling of catalytic properties.

7.4.2 Monte Carlo Simulation Tools

Monte Carlo (MC) simulation is a statistical technique widely used for the probabilistic analysis of engineering systems as well as in computational physics, physical chemistry, computational biology and related applied fields such as management. This technique depends on the random sampling for the numerical results. A typical MC simulation randomly calculates the given model numerous number of times starting from different randomly selected values. MC can be used as a substitute for computational molecular dynamics to compute statistical field theories of simple particles and polymer systems [55]. When using this technique, temperature (T), volume (V) and chemical potential (μ) of each species are fixed, while the number of gas molecules (N) in the adsorbed phase can fluctuate. Then, after equilibrium is reached, the number of molecules adsorbed is calculated using a statistically averaged approach. Chemical potential can be related to the bulk pressure or fugacity through an equation of state (EOS) such as the Peng–Robinson EOS or via a separate molecular simulation of the bulk phase. Various types of force fields have been used in MC to estimate the Lennard-Jones parameters for individual atoms of the framework [5]. Adsorption of CO_2 to different MOFs was studied using grand canonical Monte Carlo simulations. CO_2 molecules were derived by *ab initio* calculations. First, geometries of MIL MOFs (MIL-53lp (Al), MIL-53np (Al), MIL-47 (V)) were optimized using crystallographic coordinates, and then partial charges for these porous materials were extracted by DFT calculations. The Accelrys DMol^3 code and PW91 GGA density functional were used for all calculations. All calculations were performed at 303 K with typical $3 \cdot 10^6$ Monte Carlo steps [56]. Adsorption of chiral alcohols to a homochiral MOF (hybrid organic-inorganic zeolite analog (HOIZA-1)) was studied using grand Monte Carlo simulation. The pore size distribution with a pore diameter around 4.5 Å was calculated geometrically. LJ parameters for HOIZA-1 were determined from the universal force field, while the all-atom (AA) force field was used for optimized potentials liquid simulations (OPLS). Different parameters were used for (1) grid method calculations, (2) density functional theory calculations with the B3LYP functional, and (3) derivation of the LanL2DZ basis set. The grid method for charges from electrostatic potentials (CHELPG) and LanL2DZ basis set GCMC simulations were performed under liquid-phase conditions at 1-100 bar and a temperature of 300 K [57]. The validation of the transferability of simulation parameters and the local composition effects in a model MOF for adsorption of polar gases (H_2S and H_2O) and nonpolar gases (CH_4, and C_2H_6) were studied that is relevant for sensor applications [58]. A continuous fractional component Monte Carlo method was developed to overcome difficulties regarding insertions and deletions of molecules during the study of absorption on MOFs and applied to binary systems [19]. MC technique in the framework of grand canonical ensemble was used to study the adsorption equilibrium of CH_4 in PCN-14 and temperature dependence of the molecular siting using force-field method [19], which showed a surprising barrier less site behavior between weak and strong sites. The weakly adsorbing open metal clusters in PCN-14 is densely populated at low temperatures which successfully explained discrepancies between neutron diffraction experiments and MC simulations.

In MC methods, a series of points in phase space is generated from an initial geometry by adding an arbitrary unit to the coordinates of a randomly chosen particle (atom or molecule). The new configuration is accepted if the energy decreases with a probability of $e^{-\Delta E/kT}$ [52]. MC methods are essentially non-deterministic, as each configuration only depends on the previous point and a few random numbers; two simulations starting from the same geometry will not produce the same sampling because the random numbers will be different. MC simulations require only the ability to estimate the energy of the system, which can be advantageous if calculation of the first derivative is difficult or time-consuming. MC methods are, therefore, best for exploring the translational and rotational spaces for comparatively small molecules such as solvent or solution molecules as well as for determining the internal degrees of freedom of small molecules.

7.5 Outlook

Some characteristics/features of MOFs and molecules can be estimated from DFT calculations already used for diverse applications of MOFs including gas storage, separation, sensing, catalysis, adsorption, etc. Among others, the most important ones may be the absorptive, electrical and optical properties. Adsorption of molecules to open metal sites and/or on ligands provides doping of MOFs by electrons or holes. The type and value of doping will be different for different pairs of molecules and MOFs, which can be measured experimentally or obtained from DFT calculations of charge transfer [34]. Adsorption of molecules on surfaces results in changes in the vibrational properties of the substrate, which can be detected using spectroscopy or calculated by DFT [59,60]. The intramolecular vibration value is unique for each molecule and is very sensitive to the mechanism of adsorption (Figure 7.3) [45].

The structural stability of MOFs is also affected by adsorption of molecules and it was investigated by frozen phonon method in the literature [61,62]. Changes in the elastic properties of MOFs can be evaluated by comparing the total energy of pristine and strained MOFs with and without adsorbed molecules. The intramolecular vibrational frequencies can be calculated by using the following equation:

$$v = \frac{1}{\pi}\sqrt{\Delta E/2m'x^2}$$

where m' is the reduced mass, and ΔE is the difference between the total energies of the system before and after stretching of the intramolecular bond at value x.

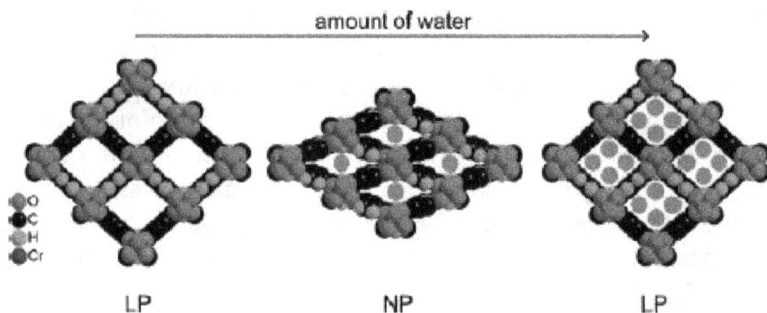

Figure 7.3 Evolution of the structure of MIL-53(Cr) upon adsorption of water (LP = large pore form, NP = narrow pore). Reproduced from Reference 45 with permission from American Chemical Society.

MOF-molecule interactions can also be characterized by changes in the magnetic properties of open metal sites. Adsorption of the molecules to open metal sites can influence magnetic moments and sometimes switch metallic ions between low and high spin configurations, as was shown for copper-based molecular magnets [63]. DFT-based calculations have been used to describe and predict changes in magnetic ground states caused by changes in the coordination of metallic ions.

The computational tools can be used to understand the basic principles underlying MOF sensing and to develop criteria to evaluate these properties. Performing DFT or MD calculations for all MOFs that have already been synthesized is not feasible. However, knowledge to predict the sensing properties of real and hypothetical MOFs can be obtained by using machine learning methods. Using these methods, systems can be characterized by a set of numerical values (descriptors) obtained from experimental measurements of calculations. Then, with the aid of various algorithms, correlation-based relations can be established between the descriptors and the properties of the studied systems. As the next step, established correlations can be used to predict the properties of real and hypothetic systems. Fortunately, for the characterization of MOFs and adsorbed molecules, organic molecules descriptors can be adopted from quantitative structure-activity relationship (QSAR) methods. These methods are successfully used in organic chemistry to design drugs and other chemicals. Adsorbed molecules and linkers of MOFs can be characterized by an existing set of descriptors used for organic molecules. Metal-ligand cores of MOFs can be characterized by sets of descriptors developed to predict the properties of inorganic compounds. Some additional values, such as the size of pores and the type and quantity of solvents, can also be considered. A machine learning approach has already been used to evaluate gas storage and separation of MOFs [12]. Further collection of experimental and theoretical data should allow the extension of these approaches to the prediction of the diverse properties of MOFs so that the design of novel effective applied materials can be facilitated.

Acknowledgements

This study was supported by a grant from the National Research Foundation of Korea (NRF) funded by the Ministry of Science, ICT & Future Planning (No. 2016R1E1A1A01940995). This work was also carried out with the support of the "Cooperative Research Program for Agriculture Science and Technology Development (Project No. PJ012521032017)", Rural Development Administration, Republic of Korea. Dr. Pawan Kumar and Dr. Tapta Kanchan Roy would like to thank UGC, New Delhi, for the 'BSR-UGC Start up Research Grant.'

Nomenclature

BTC	benzene-1,3,5-tricarboxylate
CHELPG	charges from electrostatic potentials
DFT	density functional theory
EPR	electron paramagnetic resonance
GCMC	grand canonical Monte Carlo
GGA	generalized gradient approximation
HOIZA-1	hybrid organic-inorganic zeolite analog
HOMO	highest energy occupied molecular orbital
LDA	local-density approximations

LUMO	lowest energy unoccupied molecular orbital
MM2	molecular mechanics
MC	monte Carlo
MIL	materials Institute Lavoisier
MOFs	metal organic frameworks
NPs	nanoparticles
NMR	nuclear magnetic resonance
OPLS	optimized potentials liquid simulations
PSM	post-synthesis modification
QSAR	quantitative structure-activity relationship

References

1. Alder, B. J., and Wainwright, T. E. (1959) Studies in molecular dynamics. I. General method. *The Journal of Chemical Physics*, **31**, 459-466.
2. Al-Jadir, T. M., and Siperstein, F. R. (2016) Monte Carlo simulation of adsorption of polar and nonpolar gases in (FP) YEu metal-organic framework. *Journal of Chemical & Engineering Data*, **61**, 4209-4214.
3. Baeurle, S. A. (2009) Multiscale modeling of polymer materials using field-theoretic methodologies: a survey about recent developments. *Journal of Mathematical Chemistry*, **46**, 363-426.
4. Becke, A. D. (1988) Density-functional exchange-energy approximation with correct asymptotic behavior. *Physical Review A*, **38**, 3098.
5. Campbell, J., and Tokay, B. (2017) Controlling the size and shape of Mg-MOF-74 crystals to optimise film synthesis on alumina substrates. *Microporous and Mesoporous Materials*, **251**, 190-199.
6. Chen, B., Ockwig, N. W., Millward, A. R., Contreras, D. S., Yaghi, O. M. (2005) High H_2 adsorption in a microporous metal–organic framework with open metal sites. *Angewandte Chemie*, **117**, 4823-4827.
7. Chen, B., Wang, L., Zapata, F., Qian, G., and Lobkovsky, E. B. (2008) A luminescent microporous metal–organic framework for the recognition and sensing of anions. *Journal of the American Chemical Society*, **130**, 6718-6719.
8. Chen, S., Ho, M.-H., Bullock, R. M., DuBois, D. L., Dupuis, M., Rousseau, R., and Raugel, S. (2013) Computing free energy landscapes: Application to Ni-based electrocatalysts with pendant amines for H_2 production and oxidation. *ACS Catalysis*, **4**, 229-242.
9. Colon, Y. J., and Snurr, R. Q. (2014) High-throughput computational screening of metal-organic frameworks. *Chemical Society Reviews*, **43**, 5735-5749.
10. Coudert, F.-X., and Fuchs, A. H. (2016) Computational characteriza-

tion and prediction of metal–organic framework properties. *Coordination Chemistry Reviews*, **307**, 211-236.

11. Dubbeldam, D., and Snurr, R. Q. (2007) Recent developments in the molecular modeling of diffusion in nanoporous materials. *Molecular Simulation*, **33**, 305-325.

12. Duren, T., Bae, Y.-S., and Snurr, R. Q. (2009) Using molecular simulation to characterise metal-organic frameworks for adsorption applications. *Chemical Society Reviews*, **38**, 1237-1247.

13. Wang, Y., Ge, H., Wu, Y., Ye, G., Chen, H., and Hu, X. (2014) Construction of an electrochemical sensor based on amino-functionalized metal-organic frameworks for differential pulse anodic stripping voltammetric determination of lead. *Talanta*, **129**, 100-105.

14. Kumar, P., Kumar, P., Deep, A., and Bharadwaj, L. M. (2012) Doped zinc-organic framework for sensing of pesticide. *Advanced Materials Research*, **488-489**, 1543-1546.

15. Fourches, D., Pu, D., Tassa, C., Weissleder, R., Shaw, S. Y., Mumper, R. J., and Tropsha, A. (2010) Quantitative nanostructure – activity relationship modelling. *ACS Nano*, **4**, 5703-5712.

16. Getman, R. B., Bae, Y.-S., Wilmer, C. E., and Snurr, R. Q. (2011) Review and analysis of molecular simulations of methane, hydrogen, and acetylene storage in metal–organic frameworks. *Chemical Reviews*, **112**, 703-723.

17. Giannozzi, P., Baroni, S., Bonini, N., Calandra, M., Car, R., Cavazzoni, C., Ceresoli, D., Chiarotti, G. L., Cococcioni, M., Dabo, I., Corso, A. D., de Gironcoli, S., Fabris, S., Fratesi, G., Gebauer, R., Gerstmann, U., Gougoussis, C., Kokalj, A., Lazzeri, M., Martin-Samos, L., Marzari, N., Mauri, F., Mazzarello, R., Paolini, S., Pasquarello, A., Paulatto, L., Sbraccia, C., Scandolo, S., Sclauzero, G., Seitsonen, A. P., Smogunov, A., Umari, P., and Wentzcovitch, R. M. (2009) QUANTUM ESPRESSO: a modular and open-source software project for quantum simulations of materials. *Journal of Physics: Condensed Matter*, **21**, 395502.

18. Gonze, X., Beuken, J.-M., Caracas, R., Detraux, F., Fuchs, M., Rignanese, G.-M., Sindic, L., Verstraete, M., Zerah, G., Jollet, F., Torrent, M., Roy, A., Mikami, M., Ghosez, P., Raty, J.-Y., and Allan, D. C. (2002) First-principles computation of material properties: the ABINIT software project. *Computational Materials Science*, **25**, 478-492.

19. (a) Greathouse, J. A., and Allendorf, M. D. (2006) The interaction of water with MOF-5 simulated by molecular dynamics. *Journal of the American Chemical Society*, **128**, 10678-10679; (b) Wu, L., Xiao, J., Wu, Y., Xian, S., Miao, G., Wang, H.. and Li, Z. (2014) A combined experimental/computational study on the adsorption of organosulfur compounds over metal-organic frameworks from fuels. *Langmuir*, **30**, 1080-1088.

20. Grimme, S. (2006) Semi-empirical GGA-type density functional con-

structed with a long-range dispersion correction. *Journal of Computational Chemistry*, **27**, 1787-1799.

21. Grimme, S. (2011) Density functional theory with London dispersion corrections. *Wiley Interdisciplinary Reviews: Computational Molecular Science*, **1**, 211-228.

22. Han, M. S., and Kim, D. H. (2002) Naked-eye detection of phosphate ions in water at physiological pH: A remarkably selective and easy-to-assemble colorimetric phosphate-sensing probe. *Angewandte Chemie*, **114**, 3963-3965.

23. Han, S., Kim, H., Kim, J., and Jung, Y. (2015) Modulating the magnetic behavior of Fe (ii)–MOF-74 by the high electron affinity of the guest molecule. *Physical Chemistry Chemical Physics*, **17**,16977-16982.

24. Heyd, J., Scuseria, G. E., and Ernzerhof, M. (2006) Erratum: Hybrid functionals based on a screened Coulomb potential (Journal of Chemical Physics, **118**, 8207 (2003)). *The Journal of Chemical Physics*, **124**, 219906.

25. Hohenberg, P., and Kohn, W. (1964) Inhomogeneous electron gas. *Physical Review*, **136**, B864.

26. Hu, Z., Deibert, B. J., and Li, J. (2014) Luminescent metal–organic frameworks for chemical sensing and explosive detection. *Chemical Society Reviews*, **43**, 5815-5840.

27. Israelachvili, J. N. (1992) *Intermolecular and Surface Forces*, Academic Press, USA.

28. Jorge, M., Fischer, M., Gomes, J. R. B., Siquet, C., Santos, J. C., and Rodrigues, A. E. (2014) Accurate model for predicting adsorption of olefins and paraffins on MOFs with open metal sites. *Industrial & Engineering Chemistry Research*, **53**, 15475-15487.

29. Kampert, E., Janssen, F. F. B. J., Boukhvalov, D. W., Russcher, J. C., Smits, J. M. M., de Gelder, R., de Bruin, B., Christinen, P. C. M., Zeitler, U., Katsnelson, M. I., Maan, J. C., and Rowan, A. E. (2009) Ligand-controlled magnetic interactions in Mn4 clusters. *Inorganic Chemistry*, **48**, 11903-11908.

30. Kreno, L. E., Leong, K., Farha, O. K., Allendorf, M., Van Duyne, R. P., and Hupp, J. T. (2012) Metal-organic framework materials as chemical sensors. *Chemical Reviews*, **112**, 1105-1125.

31. Kresse, G., and Furthmuller, J. (1996) Efficient iterative schemes for ab initio total-energy calculations using a plane-wave basis set. *Physical Review B*, **54**, 11169.

32. Kunc, K., and Martin, R. M. (1981) Atomic structure and properties of polar Ge-GaAs (100) interfaces. *Physical Review B*, **24**, 3445.

33. Lee, D. Y., Lim, I., Shin, C. Y., Patil, S. A., Lee, W., Shrestha, N. K., Lee, J. K., and Han, S.-H. (2015) Facile interfacial charge transfer across hole doped cobalt-based MOFs/TiO$_2$ nano-hybrids making MOFs light harvesting active layers in solar cells. *Journal of Materials Chemistry A*, **3**, 22669-22676.

34. Lee, S. J., Cho, S.-H., Mulfort, K. L., Tiede, D. M., Hupp, J. T., and Nguyen, S.-B. T. (2008) Cavity-tailored, self-sorting supramolecular catalytic boxes for selective oxidation. *Journal of the American Chemical Society*, **130**, 16828-16829.

35. Lucena, S. M., Mileo, P. G., Silvino, P. F., and Cavalcante. Jr., C .L. (2011) Unusual adsorption site behavior in PCN-14 metal-organic framework predicted from Monte Carlo simulation. *Journal of the American Chemical Society*, **133**, 19282-19285.

36. Mishra, P., Mekala, S., Dreisbach, F., Mandal, B., and Gumma, S. (2012) Adsorption of CO_2, CO, CH_4 and N_2 on a zinc based metal organic framework. *Separation and Purification Technology*, **94**, 124-130.

37. Mokrousov, Y., Bihlmayer, G., and Blugel, S. (2005) Full-potential linearized augmented plane-wave method for one-dimensional systems: Gold nanowire and iron monowires in a gold tube. *Physical Review B*, **72**, 045402.

38. Morin, C., Simon, D., and Sautet, P. (2006) Intermediates in the hydrogenation of benzene to cyclohexene on Pt (111) and Pd (111): a comparison from DFT calculations. *Surface Science*, **600**, 1339-1350.

39. Odoh, S. O., Cramer, C. J., Truhlar, D. G., and Gagliardi, L. (2015) Quantum-chemical characterization of the properties and reactivities of metal-organic frameworks. *Chemical Reviews*, **115**, 6051-6111.

40. Perdew, J. P., Burke, K., and Ernzerhof, M. (1996) Generalized gradient approximation made simple. *Physical Review Letters*, **77**, 3865-3868.

41. Politano, A., Chiarello, G., Samnakay, R., Liu, G., Gurbulak, B., Duman, S., Balandin, A. A., and Boukhvalov, D. W. (2016) The influence of chemical reactivity of surface defects on ambient-stable InSe-based nanodevices. *Nanoscale*, **8**, 8474-8479.

42. Puzyn, T., Mostrag-Szlichtyng, A., Gajewicz, A., Skrzynki, M., and Worth, A. P. (2011) Investigating the influence of data splitting on the predictive ability of QSAR/QSPR models. *Structural Chemistry*, **22**, 795-804.

43. Qiao, Z., Torres-Knoop, A., Dubbeldam, D., Fairen-Jimenez, D., Zhou, J., and Snurr, R. Q. (2014) Advanced Monte Carlo simulations of the adsorption of chiral alcohols in a homochiral metal-organic framework. *AIChE Journal*, **60**, 2324-2334.

44. Ramsahye, N. A., Maurin, G., Bourrelly, S., Liewellyn, P. L., Devic, T., Serre, C., Loiseau, T., and Ferey, G. (2007) Adsorption of CO_2 in metal organic frameworks of different metal centres: Grand Canonical Monte Carlo simulations compared to experiments. *Adsorption*, **13**, 461-467.

45. Salles, F., Bourrelly, S., Jobic, H., Devic, T., Guillerm, V., Llewellyn, P.,

Serre, C., Ferey, G., and Maurin, G. (2011) Molecular insight into the adsorption and diffusion of water in the versatile hydrophilic/hydrophobic flexible MIL-53 (Cr) MOF. *The Journal of Physical Chemistry C*, **115**, 10764-10776.

46. Schwarz, K., and Blaha, P. (2003) Solid state calculations using WIEN2k. *Computational Materials Science*, **28**, 259-273.

47. Schwerdtfeger, P. (2011) The pseudopotential approximation in electronic structure theory. *ChemPhysChem*, **12**, 3143-3155.

48. Segall, M. D., Lindan, P. J. D., Probert, M. J., Pickard, C. J., Hasnip, P. J., Clark, S. J., and Payne, M. C. (2002) First-principles simulation: ideas, illustrations and the CASTEP code. *Journal of Physics: Condensed Matter*, **14**, 2717.

49. Shvachko, Y. N., Starichenko, D., V. Korolev, A. V., Boukhvalov, D. W., and Ustinov, V. V. (2009) Interplay between π-stacking and AFM interaction in the novel triazine based dimeric crystal. *Solid State Communications*, **149**(47-48), 2189-2193.

50. Skoulidas, A. I., and Sholl, D. S. (2005) Self-diffusion and transport diffusion of light gases in metal-organic framework materials assessed using molecular dynamics simulations. *The Journal of Physical Chemistry B*, **109**, 15760-15768.

51. Smith, S. J., and Sutcliffe, B. T. (1997) The development of computational chemistry in the United Kingdom. In: *Reviews in Computational Chemistry,* Lipkowitz, K. B., and Boyd. D. B. (eds.), Wiley, USA, doi: 10.1002/9780470125878.ch5.

52. Soler, J. M., Artacho, E., Gale, J. D., Garcia, A., Junquera, J., Ordejon, P., and Sanchez-Portal, D. (2002) The SIESTA method for ab initio order-N materials simulation. *Journal of Physics: Condensed Matter*, **14**, 2745.

53. Stamenkovic, V. R., Mun, B. S., Mayrhofer, K. J. J., Ross, P. N., and Markovic, N. M. (2006) Effect of surface composition on electronic structure, stability, and electrocatalytic properties of Pt-transition metal alloys: Pt-skin versus Pt-skeleton surfaces. *Journal of the American Chemical Society*, **128**, 8813-8819.

54. Sun, F., Yin, Z., Wang, Q.-Q., Sun, D., Zeng, M.-H., and Kurmoo, M. (2013) Tandem post-synthetic modification of a metal–organic framework by thermal elimination and subsequent bromination: effects on absorption properties and photoluminescence. *Angewandte Chemie*, **125**(17), 4636-4641.

55. Van Schilfgaarde, M., Kotani, T., and Faleev, S. (2006) Quasiparticle self-consistent G W theory. *Physical Review Letters*, **96**, 226402.

56. Vellingiri, K., Deep, A., Kim, K.-H., Boukhvalov, D. W., Kumar, P., and Yao, Q. (2017) The sensitive detection of formaldehyde in aqueous media using zirconium-based metal organic frameworks. *Sensors and Actuators B: Chemical*, **241**, 938-948.

57. Vellingiri, K., Szulejko, J. E., Kumar, P., Kwon, E. E., Kim, K.-H., Deep,

A., Boukhvalov, D. W., and Brown, R. J. (2016) Metal organic frameworks as sorption media for volatile and semi-volatile organic compounds at ambient conditions. *Scientific Reports*, **6**, 27813.

58. Fischer, C. F., Lagowski, J. B., and Vosko, S. (1987) Ground states of Ca– and Sc– from two theoretical points of view. *Physical Review Letters*, **59**, 2263.

59. Xiao, W., Hu, C., Carter, D. J., Nichols, S., Ward, M. D., Raiteri, P., Rohl, A. L., and Kahr, B. (2014) Structural correspondence of solution, liquid crystal, and crystalline phases of the chromonic mesogen sunset yellow. *Crystal Growth and Design*, **14**, 4166-76.

60. Xu, H., Liu, F., Cui, Y., Chen, B., and Qian, G. (2011) A luminescent nanoscale metal-organic framework for sensing of nitroaromatic explosives. *Chemical Communications*, **47**, 3153-3155.

61. Yang, Q., Bu, X., Zhong, C., and Li, Y.-g. (2005) Molecular simulation of adsorption of HCFC-22 in pillared clays. *AIChE Journal*, **51**, 281-291.

62. Yang, Q., Vaesen, S., Ragon, F., Wiersum, A. D., Wu, D., Lago, A., Devic, T., Martineau, C., Taulelle, F., Llewellyn, P. L., Jobic, H., Zhong, C., Serre, C., De Weireld, G., and Maurin, G. (2013) A water stable metal-organic framework with optimal features for CO_2 capture. *Angewandte Chemie International Edition*, **52**, 10316-10320.

63. Zhu, C., Wu, J., Li, S., Yang, Y., Zhu, J., Lu, X., and Xia, H. (2017) Synthesis and characterization of a metallacyclic framework with three fused five-membered rings. *Angewandte Chemie International Edition*, **56**(31), 9067-9071.

8

Metal Organic Frameworks for H₂S Separation from Natural Gas

Gigi George

Department of Chemistry, CMS College Kottayam, Mahatma Gandhi University, Kerala 686001, India

gigi@cmscollege.ac.in

8.1 Introduction

Hydrogen sulfide (H₂S) is one of the most significant impurities in natural gas streams. Its presence can lead to corrosion, reduction in the thermal content of the natural gas and formation of harmful hydrides [1-3]. Due to these effects, its removal from natural gas is carried out using different methodologies including sweetening, absorption by amines and adsorption. Adsorption of H₂S has been reported on various types of adsorbents, ranging from activated carbons (ACs) [4-6] to zeolites [6,7] and metal organic frameworks (MOFs) [8-10]. Each type of adsorbent exhibits specific advantages and limitations attributed to its chemical and physical structure, manufacturing methods and cost. Figure 8.1 also exhibits the basic structural differences among crystal structures of different MOFs used for the separation of H₂S from natural gas.

MOFs, discovered in 1989 [11], consist of metal ions or clusters joined together by organic linkers via covalent interactions and form 1D, 2D or 3D infinite networks [12]. Most commonly used metal ions in MOFs include Cu^+, Cu^{2+}, Li^+, Mg^{2+}, Cd^{2+}, Ag^+, Zn^{2+} and Co^{2+}. The common organic linkers include imidazolates, polycarboxylates, sulfonates, amines, phosphonates and phenolates [13]. MOFs present specific advantages when compared to the other classes of adsorbents; they are highly porous, and their porosity can be tailored in the micro- and mesoporous regions. Their surface areas can reach up to 10,000 m² g⁻¹, and their metal sites offer specific interaction potential towards adsorbate molecules. MOFs are theoretically

Metal Organic Frameworks, edited by Vikas Mittal
© 2019 Central West Publishing, Australia

infinite, since, in contrast to inorganic crystals, most metals can be used for the preparation of MOFs in combination with ligands differing in nature and length, thus, leading to the analogous species with varying porosities and structures [14,15].

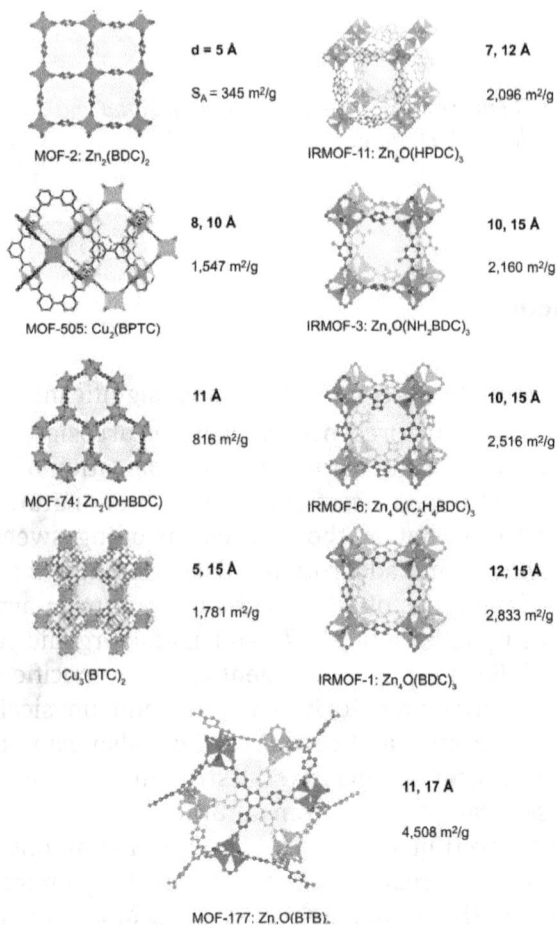

Figure 8.1 Various crystal structures of MOFs. Reproduced from Reference 8 with permission from American Chemical Society.

Besides these advantages, MOFs have significant shortcomings that become more pronounced in the case of H_2S adsorption. Such shortcomings include poor stability under humid conditions and elevated temperatures, along with weak dispersion forces due to their low specific densities [16]. MOFs based on metals in 2+, 3+ and 4+ states have been proved to exhibit high resistance against humidity, high sulfur selectivity and extreme resistance towards corrosive gases. These MOFs can also be regenerated using moderate conditions [17]. Specific density has been shown to be the primary factor affecting the strength of dispersion forces, especially in low temperature ranges. Adsorbents exhibiting strong electrostatic forces (like MOFs) tend to adsorb small molecules in their structures in greater degrees as compared to hydrophobic adsorbents like activated carbons; the latter adsorb small molecules only by dispersion (van der Waals forces).

8.2 Adsorption Mechanisms

Adsorption on MOFs is suggested to occur by five different mechanisms, as presented in the study by Ma *et al.* [18] on the effect of H_2S on living cells: i) acid-base interactions [19], ii) coordination bonding [20,21], iii) π-complexation [15,22,23], iv) selective absorption of hydrogen sulfide from gas streams [9,24] and v) van der Waals interactions (dispersion forces) [25]. Figure 8.2 exhibits the various

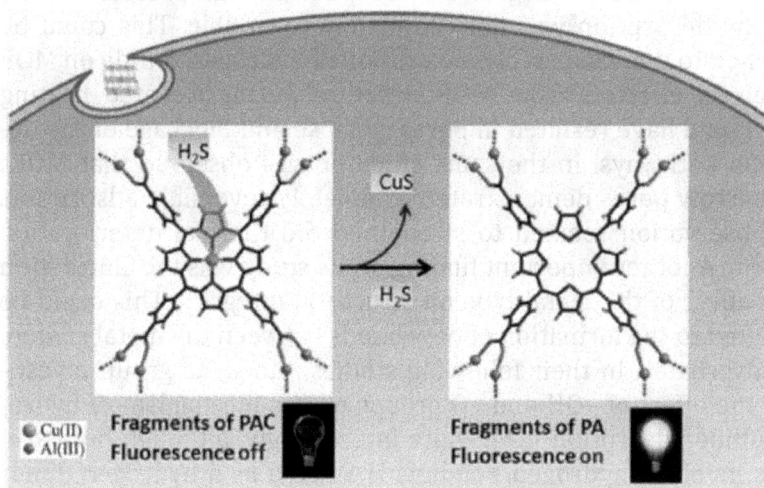

Figure 8.2 Adsorption mechanisms MOFs in living cells. Reproduced from Reference 18 with permission from American Chemical Society.

adsorption mechanisms. Adsorption of gaseous species on MOFs can be attributed to more than one mechanism. Acid-base interaction is one of the most predominant mechanisms for the capture of contaminants by MOFs; acidic gases are captured by basic MOFs and vice versa. Acidic and/or basic character of MOFs can be strength-ened further by grafting of functional groups on metal sites, utiliza-tion of fillers and use of linkers carrying functional groups [15-23]. Π- complexation is observed in specific MOFs and adsorbates only, as it requires the presence of a d-block metal for the adsorption to take place. This mechanism proceeds via donation of π-electrons from the adsorbate to the s-orbital of the metal site and a subse-quent donation of electrons from a metal's d- orbital to π* of the ad-sorbate. Coordination bond formation requires the presence of a coordination unsaturated site (metal), and it mainly involves specif-ic acid-base interactions. Hydrogen bonding occurs when polar bonds exist in an adsorbent which can bind the contaminants.

8.3 H_2S Adsorption on MOFs

H_2S adsorption at room temperature was investigated on a group of popular MOFs, including MIL-47 (V), MIL-53 (Al, Cr and Fe), MIL-100 (Cr) and MIL-101 (Cr) [9]. The study revealed that the larger pores led to a greater degree of adsorption at high pressure. How-ever, the desorption was not completely reversible. This could be attributed to two factors: chemisorption of hydrogen sulfide on MOF or thermal effects on the MOF structure during elevated heating which could have resulted in pores collapse and blockage of the de-sorption pathways. In the same study, it was observed that MOFs with narrow pores demonstrated completely reversible adsorption. This observation seemed to strengthen the thermal deterioration scenario. Another important finding of the study was the elucidation of the effect of the metal type on structural integrity. This could be attributed to the formation of new bonds between the metal centers and adsorbates. In their following studies, the same group investi-gated the effect of –OH and –O groups on the mechanism of hydro-gen sulfide adsorption [26]. Quite interestingly, although both sce-narios involved hydrogen bonding, H_2S acted as a hydrogen donor when adsorbed on MIL-47 (V) and as an acceptor on MIL-53 (Cr). The adsorption isotherms on both MOFs could be described via condensation in the pores.

MIL-53 (Al) has been commonly used for the adsorption of H_2S and other gases [27]. Although a specific affinity was observed to-wards hydrogen sulfide, however, no further investigation on the adsorption mechanism was reported. Other studies focused on the examination of graphene or graphitic oxide (GO)/HKUST-1 (Cu) composites [28]. Incorporation of graphene sheets led to an obvious increase in H_2S adsorption, in comparison to pristine MOF and GO. This effect was attributed to the 'improvement' of pore network, though more than just re-ordering of pore sizes and shapes must have occurred. It is quite possible that three bodies' interactions were generated in this case, leading to optimum coordination or π-complexation bonding between the adsorbate and composite adsorbent. Adsorption was characterized as chemisorption (reactive adsorption) which strengthened this hypothesis. Similar adsorption enhancement was reported for MOF-5/GO composite for an optimum 5% concentration of GO [29]. Schematic view of the MOF structural unit is presented in Figure 8.3. In all of the mentioned cases, the enhancement in the dispersive forces was significant due to an increase in the solid density, and it was claimed to be the sole factor for the observed enhancement in adsorption.

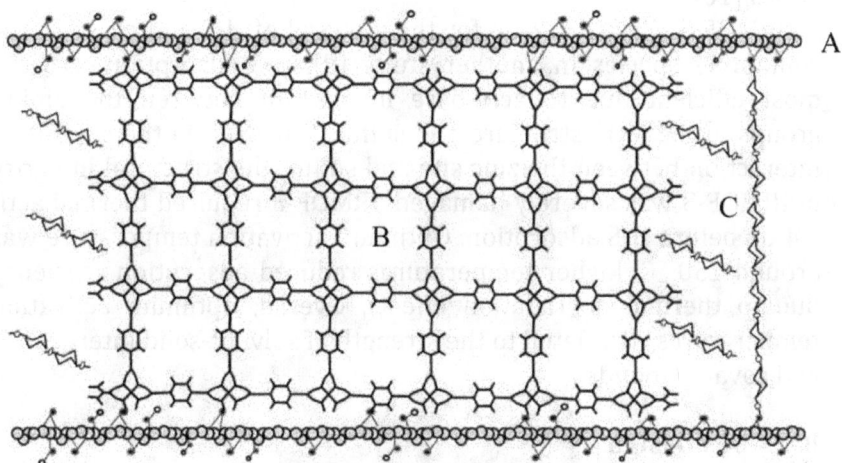

Figure 8.3 ((A) GO layer, (B) MOF-5, and (C) glucose polymer. Reproduced from Reference 29 with permission from American Chemical Society.

HKUST-1 and ZIF-8 have also been investigated as H_2S adsorbents [17]. Different techniques (FT-IR, Raman and PXRD) were utilized to monitor the structural changes during H_2S adsorption. It was observed that the changes were initiated even at very low partial pressures, indicating interaction between the metal sites and sulfur. These interactions were more pronounced at higher pressures (20-60 mbar), leading to complete collapse of MOF structures. In such cases, functionalization of ligands is advised to drive the interactions towards acid-base mechanism involving functional groups (and not metal sites). Another study discussed the issue of HKUST-1 deterioration based on the H_2S and H_2O adsorption energies (-43.4 and -46.7 kJ mol^{-1}, respectively). The observed energy values indicated a gap in the suggested adsorption mechanism [30]. This view is further supported by the theoretical examination of the adsorption energies of H_2O and H_2S on CPO-27-Ni [31].

Reduced H_2S adsorption was observed on MIL-68 (Al) in another study, and it was reported that the triangular shape of pores did not allow the adsorption process [32]. The pores were easily blocked by contaminants, thus, requiring activation. The theoretical investigation revealed that the activated pore structure allows for superb H_2S adsorption after complete removal of contaminants. Another study also reported the adsorption on Zn based MOF (MOF-74), which exhibited perfect structure integrity through adsorption desorption cycles [10].

IRMOF-3 (Zn) was used for the removal of H_2S and other sulfur containing species in another study [8]. H_2S adsorption was the most efficient due to acid-base interaction between the amino groups in the MOF structure and sulfur. However, in the case of an interaction between the zinc site and sulfur, the structural integrity of IRMOF-3 was severely damaged. IRMOF-3 required thermal activation before H_2S adsorption. Optimum activation temperature was around 150 °C; higher temperatures reduced adsorption efficiency due to thermal degradation effects. Overall, optimum activation temperatures correlated to the strength of solvent-solid interactions and covalent bonds.

8.4 MOFs Design

Another approach to construct MOFs resistive towards H_2S destabilization was reported on the basis of rare earth metals. A series of isomorphic porous MOFs was synthesized based on the hydrothermal methodology for this reason, having the general formula $Re_2(phen)_2(fdc)_2 \cdot 2H_2O$, where Re stands for Y, Eu and $Y_{1.9}Eu_{0.1}$. Water removal and activation took place around 200 °C with no evidence of destabilization. The specific surface area of MOFs was low (approx.. 172 m^2 g^{-1}), which validated reactive adsorption of H_2S even at low pressures. The most probable interactions between the adsorbate and MOF seemed to be acid-base (without the participation of the metal site) or donor-acceptor types [33]. Figure 8.4

demonstrates the coordination modes of MOFs with aluminum, chromium and iron in the framework, investigated for hydrogen sulfide adsorption at room temperature [9]. Another possible mechanism included the partial participation of the rare earth metal sites. For this purpose, Ln(BTC)(H$_2$O)•(DMF) was synthesized with the use of 1,3,5 trimesic acid (BTC) as a ligand and Ln=Sm, Eu, Tb, Y as rare earth metals [34]. Rare earth metals have been investigated due to their partially filled 4f electronic shell that leads to a high co-ordination number. S_{BET} for all MOFs were in the range 512-656 m^2 g^{-1}. Adsorption profiles led to the determination of mechanisms as CUS adsorption (adsorption on a coordination unsaturated site) and π- complexation.

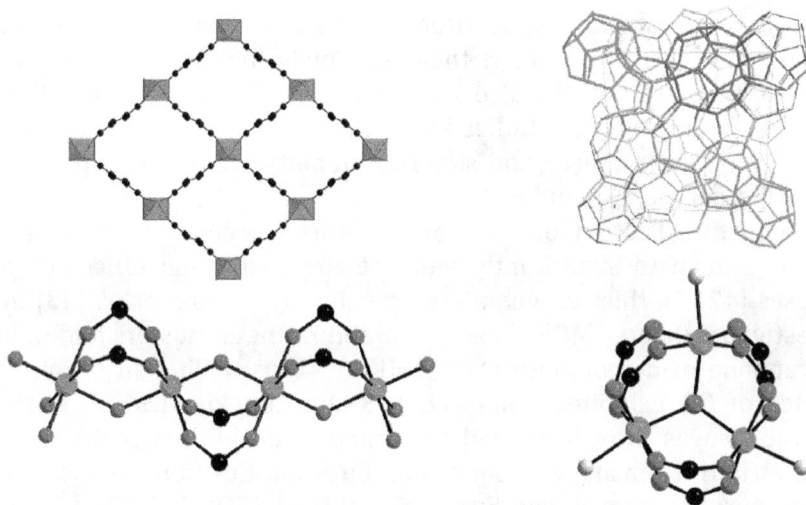

Figure 8.4 Coordination modes of fdc2- ligand, showing two types of link modes. Reproduced from Reference 9 with permission from American Chemical Society.

Another approach investigated by Chaemchuen *et al.* [35] is to create different porosities and pore sizes by using ligands or different sizes. M-DABCO series of MOFs were prepared by using two metal atoms combined with four BDC linkers and two DABCO ligands in the framework. The structural stability in the study was attributed to the coordination of the metal sites to N and O atoms as well as the absence of free sites for H$_2$O or H$_2$S attack. Also, the novel 3D structure of the M-DABCO series protected the metal sites. In addition, the benzene groups carried by the linkers are known to be hydrophobic agents [36] that contribute to the repulsion of water molecules. Various metals utilized in the study (Ni, Co, Cu, Zn) led to different local charge densities which also affected the adsorption process. These insights should be further investigated in detail via density functional theory or *ab initio* calculations to achieve optimum metal ligand sizes and combinations for H_2S adsorption and structural stability.

An alternative approach relates to the use of COFs, which do not include metal sites. In a recent study, the adsorption of H_2S on a series of COFs was investigated [37]. It was concluded that the pore volume and available surface area governed the H_2S adsorption efficiency. This finding is in good agreement with the mentioned adsorption mechanisms on MOFs. Absence of metal sites in COFs leads to different pathways of adsorbent-adsorbate interactions. The advantage of utilizing COFs is their extremely high porosity and tailored pore sizes. COFs also exhibit higher thermal stability than MOFs [38,39]. Recent studies have led to the development of COFs with high surface area and superior stability for the adsorption of small gases like CO_2 and H_2 [40,41].

Functionalization of microporous carbonaceous materials has been shown to significantly enhance the adsorption efficiency of gases [42]. In this respect, a recent study by Pokhrel *et al.* [43] investigated diverse MOFs bearing Cu-and Zr-metal clusters and their graphene oxide composites. The HKUST-1 crystallites in the presence of GO exhibited enhanced H_2S sorption kinetics. A general strategy was also developed by Zhang *et al.* [44] to realize ratiometric H_2S sensing by condensing Eu^{3+} and Cu^{2+} ions in the MOF structure. The developed Eu^{3+}/Cu^{2+}@UiO-66-$(COOH)_2$ MOF (Figure 8.5) exhibited high sensing capabilities.

8.5 Concluding Remarks and Future Perspectives

The chapter presents the current status of the utilization of MOFs for H_2S adsorption for separating it from gaseous mixtures. It has been concluded that MOFs exhibit a high affinity towards H_2S and efficient storage can be achieved at ambient and higher temperatures. The findings also show the advantages of MOFs in comparison to other porous materials such as activated carbons where the adsorption is based mainly on dispersion forces. The most significant advantage of MOFs is the unlimited potential for the modification of pore sizes and chemical structures via the use of different ligands and metal sites. The superior performance of MOFs is attributed to the different mechanisms of adsorption: i) acid base interactions, ii) coordination bonding, iii) π- complexation, iv) hydrogen bonding and v) van der Waals interactions (dispersion forces). Besides van

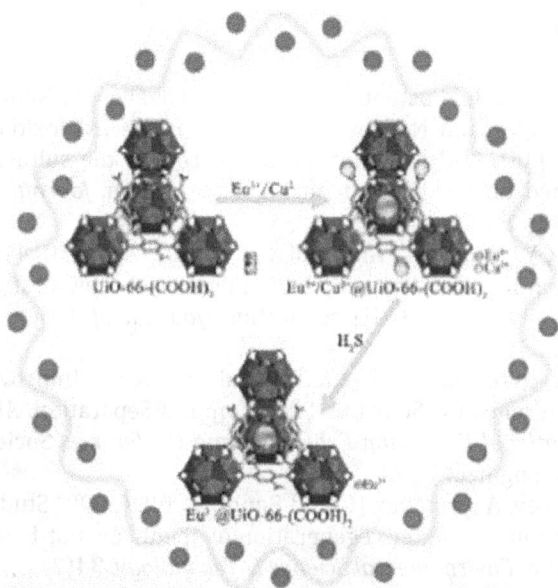

Figure 8.5 Crystalline structure of Eu^{3+}/Cu^{2+}@UiO-66-(COOH)₂ MOF. Reproduced from Reference 44 with permission from American Chemical Society.

der Waals interactions, all mechanisms are dependent on the metal site and its interactions both with the solid matrix and the adsorbate molecules/atoms. This plethora of interactions leads to a significant enhancement of adsorption potential at ambient and higher temperatures.

MOFs have high cost of manufacturing and suffer from poor stability under humid environments and high temperatures. It has been concluded that specific metal groups (rare earth metals) can be used on their own or in combination with typical metals (such as Zn, Mg, Fe, etc.) in order to construct MOFs that are both thermo-resistive and stable against sulfur compounds.

Future research should aim at exploring new manufacturing pathways in order to reduce the cost of MOFs and, at the same time, identify optimum rare earth metal combinations that exhibit superior stability under humid and high temperature conditions. Extensive theoretical investigations are also needed for achieving this objective.

References

1. Laperdrix, E., Costentin, G., Saur, O., Lavalley, J., Nedez, C., Savin-Poncet, S., and Nougayrede, J. (2000) Selective oxidation of H2S over CuO/Al$_2$O$_3$: Identification and role of the sulfurated species formed on the catalyst during the reaction. *Journal of Catalysis*, **189**(1), 63-69.
2. Zhao, Y., Jung, B. T., Ansaloni, L., and Ho, W. W. (2014) Multiwalled carbon nanotube mixed matrix membranes containing amines for high pressure CO$_2$/H$_2$ separation. *Journal of Membrane Science*, **459**, 233-243.
3. George, G., Bhoria, N., and Mittal, V. (2015) Improved Polymer Membranes for Sour Gas Filtration and Separation. *Abu Dhabi International Petroleum Exhibition and Conference*, Society of Petroleum Engineers, UAE.
4. Bagreev, A., Rahman, H., and Bandosz, T. J. (2000) Study of H2S adsorption and water regeneration of spent coconut-based activated carbon. *Environmental Science & Technology*, **34**(21), 4587-4592.
5. Bandosz, T. J. (2002) On the adsorption/oxidation of hydrogen sulfide on activated carbons at ambient temperatures. *Journal of Colloid and Interface Science*, **246**(1), 1-20.
6. Yuan, W., and Bandosz, T. J. (2007) Removal of hydrogen sulfide from biogas on sludge-derived adsorbents. *Fuel*, **86**(17-18), 2736-2746.
7. Yan, R., Chin, T., Ng, Y. L., Duan, H., Liang, D. T., and Tay, J. H. (2004) Influence of surface properties on the mechanism of H2S removal by alkaline activated carbons. *Environmental Science & Technology*, **38**(1), 316-323.
8. Wu, C.-D., Hu, A., Zhang, L., and Lin, W. (2005) A homochiral porous metal- organic framework for highly enantioselective heterogeneous asymmetric catalysis. *Journal of the American Chemical Society*, **127**(25), 8940-8941.
9. Hamon, L., Serre, C., Devic, T., Loiseau, T., Millange, F., Férey, G., and Weireld, G. D. (2009) Comparative study of hydrogen sulfide adsorption in the MIL-53 (Al, Cr, Fe), MIL-47 (V), MIL-100 (Cr), and MIL-101 (Cr) metal–organic frameworks at room temperature. *Journal of the American Chemical Society*, **131**(25), 8775-8777.
10. Allan, P. K., Wheatley, P. S., Aldous, D., Mohideen, M. I., Tang, C.,

Hriljac, J. A., Megson, I. L., Chapman, K. W., De Weireld, G., Vaesen, S., and Morris, R. E. (2012) Metal-organic frameworks for the storage and delivery of biologically active hydrogen sulfide. *Dalton Transactions*, **41**(14), 4060-4066.

11. Hoskins, B. F., and Robson, R. (1989) Infinite polymeric frameworks consisting of three dimensionally linked rod-like segments. *Journal of the American Chemical Society*, **111**(15), 5962-5964.

12. Britt, D., Tranchemontagne, D., and Yaghi, O. M. (2008) Metal-organic frameworks with high capacity and selectivity for harmful gases. *Proceedings of the National Academy of Sciences*, **105**(33), 11623-11627.

13. Horcajada, P., Gref, R., Baati, T., Allan, P. K., Maurin, G., Couvreur, P., Ferey, G., Morris, R. E., and Serre, C. (2011) Metal–organic frameworks in biomedicine. *Chemical Reviews*, **112**(2), 1232-1268.

14. Eddaoudi, M., Kim, J., Rosi, N., Vodak, D., Wachter, J., O'keeffe, M., and Yaghi, O. M. (2002) Systematic design of pore size and functionality in isoreticular MOFs and their application in methane storage. *Science*, **295**(5554), 469-472.

15. Jhung, S. H., Khan, N. A., and Hasan, Z. (2012) Analogous porous metal–organic frameworks: Synthesis, stability and application in adsorption. *CrystEngComm*, **14**(21), 7099-7109.

16. Petit, C., Levasseur, B., Mendoza, B., and Bandosz, T. J. (2012) Reactive adsorption of acidic gases on MOF/graphite oxide composites. *Microporous and Mesoporous Materials*, **154**, 107-112.

17. Ethiraj, J., Bonino, F., Lamberti, C., and Bordiga, S. (2015) H2S interaction with HKUST-1 and ZIF-8 MOFs: A multitechnique study. *Microporous and Mesoporous Materials*, **207**, 90-94.

18. Ma, Y., Su, H., Kuang, X., Li, X., Zhang, T., and Tang, B. (2014) Heterogeneous nano metal-organic framework fluorescence probe for highly selective and sensitive detection of hydrogen sulfide in living cells. *Analytical Chemistry*, **86**(22), 11459-11463.

19. Khan, N. A., and Jhung, S. H. (2012) Adsorptive removal of benzothiophene using porous copper-benzenetricarboxylate loaded with phosphotungstic acid. *Fuel Processing Technology*, **100**, 49-54.

20. Huo, S.-H., and Yan, X.-P. (2012) Metal–organic framework MIL-100 (Fe) for the adsorption of malachite green from aqueous solution. *Journal of Materials Chemistry*, **22**(15), 7449-7455.

21. Glover, T. G., Peterson, G. W., Schindler, B. J., Britt, D., and Yaghi, O. (2011) MOF-74 building unit has a direct impact on toxic gas adsorption. *Chemical Engineering Science*, **66**(2), 163-170.

22. Khan, N. A., Jung, B. K. Hasan, Z. and Jhung, S. H. (2015) Adsorption and removal of phthalic acid and diethyl phthalate from water with zeolitic imidazolate and metal-organic frameworks. *Journal of Hazardous Materials*, **282**, 194-200.

23. Khan, N. A., Hasan, Z., and Jhung, S. H. (2013) Adsorptive removal of hazardous materials using metal-organic frameworks (MOFs): a review. *Journal of Hazardous Materials*, **244**, 444-456.

24. Frazier, H. and A. Kohl. (1950) Selective absorption of hydrogen sulfide from gas streams. *Industrial & Engineering Chemistry*, 1950. **42**(11), 2288-2292

25. Hamon, L., Leclerc, H., Ghoufi, A., Oliviero, L., Travert, A., Lavalley, J.-C., Devic, T., Serre, C., Férey, G., and De Weireld, G. (2011) Molecular insight into the adsorption of H2S in the flexible MIL-53 (Cr) and rigid MIL-47 (V) MOFs: infrared spectroscopy combined to molecular simulations. *The Journal of Physical Chemistry C*, **115**(5), 2047-2056.

26. Ahmed, I., and Jhung, S. H. (2016) Adsorptive desulfurization and denitrogenation using metal-organic frameworks. *Journal of Hazardous Materials*, **301**, 259-276.

27. Heymans, N., Vaesen, S., and De Weireld, G. (2012) A complete procedure for acidic gas separation by adsorption on MIL-53 (Al). *Microporous and Mesoporous Materials*, **154**, 93-99.

28. Petit, C., Mendoza, B., and Bandosz, T. J. (2010) Hydrogen sulfide adsorption on MOFs and MOF/graphite oxide composites. *ChemPhysChem*, **11**(17), 3678-3684.

29. Huang, Z.-H., Liu, G., and Kang, F. (2012) Glucose-promoted Zn-based metal-organic framework/graphene oxide composites for hydrogen sulfide removal. *ACS Applied Materials & Interfaces*, **4**(9), 4942-4947.

30. Gutiérrez-Sevillano, J. J., Martín-Calvo, A., Dubbeldam, D., Calero, S., and Hamad, S. (2013) Adsorption of hydrogen sulphide on metal-organic frameworks. *RSC Advances*, **3**(34), 14737-14749.

31. Chavan, S., Bonino, F., Valenzano, L., Civalleri, B., Lamberti, C., Acerbi, N., Cavka, J. H., Leistner, M., and Bordiga, S. (2013) Fundamental aspects of H2S adsorption on CPO-27-Ni. *The Journal of Physical Chemistry C*, **117**(30), 15615-15622.

32. Yang, Q., Vaesen, S., Vishnuvarthan, M., Ragon, F., Serre, C., Vimont, A., Daturi, M., De Weireld, G., and Maurin, G. (2012) Probing the adsorption performance of the hybrid porous MIL-68 (Al): a synergic combination of experimental and modelling tools. *Journal of Materials Chemistry*, **22**(20), 10210-10220.

33. Shi, F.-N., Pinto, M. L., Ananias, D., and Rocha, J. (2014) Structure, topology, gas adsorption and photoluminescence of multifunctional porous RE 3+-furan-2, 5-dicarboxylate metal organic frameworks. *Microporous and Mesoporous Materials*, **188**, 172-181.

34. Xiang, L., Jingyan, W., Qingyuan, L., Jiang, S., Zhang, T., and Shengfu, J. (2014) Synthesis of rare earth metal-organic frameworks (Ln-MOFs) and their properties of adsorption desulfurization. *Journal of Rare Earths*, **32**(2), 189-194.

35. Chaemchuen, S., Zhou, K., Kabir, N. A., Chen, Y., Ke, X., Van Tendeloo, G., and Verpoort, F. (2015) Tuning metal sites of DABCO MOF for gas purification at ambient conditions. *Microporous and Mesoporous Materials*, **201**, 277-285.

36. Song, P., Li, Y., He, B., Yang, J., Zheng, J., and Li, X. (2011) Hydrogen storage properties of two pillared-layer Ni (II) metal-organic frameworks. *Microporous and Mesoporous Materials*, **142**(1), 208-213.

37. Wang, H., Zeng, X., Wang, W., and Cao, D. (2015) Selective capture of trace sulfur gas by porous covalent-organic materials. *Chemical Engineering Science*, **135**, 373-380.

38. El-Kaderi, H. M., Hunt, J. R., Mendoza-Cortés, J. L., Côté, A. P., Taylor, R. E., O'keeffe, M., and Yaghi, O. M. (2007) Designed synthesis of 3D covalent organic frameworks. *Science*, **316**(5822), 268-272.

39. Cote, A. P., El-Kaderi, H. M., Furukawa, H., Hunt, J. R., and Yaghi, O. M. (2007) Reticular synthesis of microporous and mesoporous 2D covalent organic frameworks. *Journal of the American Chemical Society*, **129**(43), 12914-12915.

40. Lan, J., Cao, D., Wang, W., and Smit, B. (2010) Doping of alkali, alkaline-earth, and transition metals in covalent-organic frameworks for enhancing CO2 capture by first-principles calculations and molecular simulations. *ACS Nano*, **4**(7), 4225-4237.

41. Lan, J., Cao, D., Wang, W., Ben, T., and Zhu, G. (2010) High-capacity hydrogen storage in porous aromatic frameworks with diamond-like structure. *The Journal of Physical Chemistry Letters*, **1**(6), 978-981.

42. Georgakis, M., Stavropoulos, G., and Sakellaropoulos, G. (2014) Alteration of graphene based slit pores and the effect on hydrogen molecular adsorption: A simulation study. *Microporous and Mesoporous Materials*, **191**, 67-73.

43. Pokhrel, J., Bhoria, N., Wu, C., Reddy, K. S. K., Margetis, H., Anastasiou, S., George, G., Mittal, V., Romanos, G., Karonis, D., and Karanikolos, G. N. (2018) Cu- and Zr-based metal organic frameworks and their composites with graphene oxide for capture of acid gases at ambient temperature. *Journal of Solid State Chemistry*, **266**, 233-243.

44. Zhang, X., Hu, Q., Xia, T., Zhang, J., Yang, Y., Cui, Y., Chen, B., and Qian, G. (2016) Turn-on and ratiometric luminescent sensing of hydrogen sulfide based on metal-organic frameworks. *ACS Applied Materials & Interfaces*, **8**(47), 32259-32265.

9

Metal Organic Frameworks for Detection of Organic Nitro Compounds

Yadagiri Rachuri,[a,b,c] Bhavesh Parmar,[b,c] Jintu Francis Kurisingal,[a]
Kamal Kumar Bisht,[d] Dae-Won Park[a] and Eringathodi Suresh[b,c]*

[a]Division of Chemical and Biomolecular Engineering, Pusan National
University, Busan 46241, Republic of Korea
[b]Academy of Scientific and Innovative Research (AcSIR), CSIR-Central Salt
and Marine Chemicals Research Institute, G. B. Marg, Bhavnagar-364 002,
Gujarat, India
[c]Analytical and Environmental Science Division & Centralized Instrument
Facility, CSIR-Central Salt and Marine Chemicals Research Institute, G. B.
Marg, Bhavnagar-364 002, India
[d]Department of Chemistry, RCU Government Post Graduate College,
Uttarkashi-249193, Uttarakhand, India

*Corresponding author: esuresh@csmcri.res.in

9.1 Introduction

Nitro organic explosives and chemical warfare agents provide a wide range of potential weapons for anti-social elements, due to their ease of producing and deploying, leading to significant loss of life and property. Consequently, many countries around the world have been paying special attention to the selective and rapid detection of chemical explosives in the areas such as homeland security, civilian safety, and environment. The recent rise in the global disturbances has encouraged the necessity of standoff/remote, sensitive and low-cost detection of explosives or explosive like compounds. In addition, given the widespread use of explosive formulations, the analysis of explosives has been of importance in forensic research, landmine detection and environmental complications associated with explosive residues [1-10]. The extensive use of explosives for military purposes has also raised concerns about environmental contamination, especially in the areas of production and storage. The organic nitro compounds (ONCs) are also widely utilized in various industries, especially TNP

Metal Organic Frameworks, edited by Vikas Mittal
© 2019 Central West Publishing, Australia

is used in the leather, pharmaceutical, fireworks, crackers and dye industries. Significant public health hazards, including anemia, carcinogenicity, abnormal liver functions, cataract development and skin irritation, etc., could also be posed to both animals and human beings through short-term or long-term exposure of explosives. For example, TNT is classified as an EPA pollutant and one of the high intensity explosives, whereas TNP is one of the most dangerous chemicals with an explosive potential even higher than TNT. TNP and other nitroaromatic/nitroaliphatic compounds such as RDX, DMNB, NM, TNG, 2,4-DNP, 2,6-DNT and tetryl are also toxic pollutants with mutagenic properties, which lead to the contamination of the soil and water bodies if released. Therefore, rapid, sensitive and selective detection of nitro aromatic compounds could provide swift warning in the case of spills and extremist attacks, help in tracking and locating explosive compounds as well as reducing the fatalities from landmines, and offer appropriate feedback during the characterization and remediation of contaminated sites. The molecular structures of nitroaromatic and nitroaliphatic based explosive molecules are depicted in Figure 9.1. Various methods for the detection of organic nitro compounds (explosives) are currently available such as gas chromatography coupled with mass spectrometry, gas chromatography-electron capture detection, surface-enhanced Raman spectroscopy, mass spectrometry, X-ray imaging, electrochemical procedures, thermal neutron analysis and ion mobility spectroscopy (IMS) [11-20].

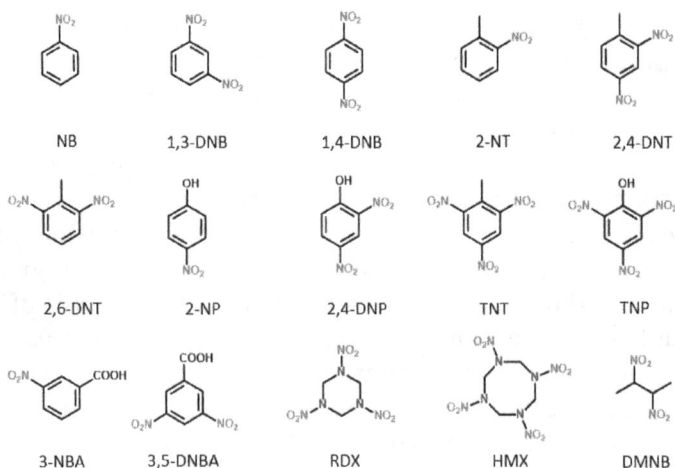

Figure 9.1 Chemical structures of NACs and nitroaliphatic based explosive and explosive like molecules.

IMS instruments, generally used in airports, have a detection sensitivity up to ppb level, but they are expensive, not easily moveable, liable to false positives and need frequent calibration [20]. Another common and effective way to detect the explosives is the use of trained canines, although the process is labor intensive and not suitable for continuous monitoring [21]. Luminescent materials have been extensively explored and utilized as lighting, display and optical devices as well as chemical sensors [22-25]. In fact, some inorganic luminescent materials, for example $BaMgAl_{10}O_{17}:Eu^{2+}$ and $GdMgB_5O_{10}:Ce^{3+},Tb^{3+}$, have been commercially employed for blue and green luminescent lamps, respectively. Luminescent CdS nanoparticles represent another remarkable material in this category, whose emitted light can be tuned from green to red simply by changing the particle size [26-28]. Recently, conjugated polymers have emerged as reliable alternatives for various applications including detection of organic nitro compounds (explosives) [9,29-30]. Due to the conjugated double bonds, they are rich in delocalized π-electrons which make them good luminescent materials for detection of targeted molecules. The electron transfer mechanism *via* donor-acceptor fluorescence quenching is the primary phenomenon for these materials [4,9]. For example, the delocalized π* excited state increases the electron donating/releasing ability to facilitate the excitation migration process in conjugated polymers based on polyacetylene [31], metalloles [32] and phenyleneethynylene [33]. This significantly enhances the electrostatic interactions between the polymeric aromatic moieties and improves their interaction with the explosive or explosive like organic nitro molecules. Analogous electronic process is also observed in fluorescent molecular species where electron transfer is the principal mechanism [34].

Owing to the potential applications, many hybrid materials have been developed by exploiting inorganic and organic components which can provide the platforms to generate luminescence. The advanced forms of self-assembly have been developed for this purpose, such as coordination polymers (CPs) or coordination networks, also popularly known as the metal organic frameworks (MOFs) [35,36]. MOFs/CPs, as indicated by the name, are crystalline solids constructed *via* self-assembly of metal cations (primary building unit or PBU) or metal clusters (secondary building unit or SBU) and organic ligands having multiple binding sites (N-donor ligands and/or multi carboxylic acids), forming one, two and three dimensional extended coordination networks [37]. Many of these crystalline solid materials are highly porous with large surface areas (up to 7,000 m²/g) and average pore diameters less than 1 nm [38]. Due to their intrinsic porous framework, with tunable pore size and structural integrity, these materials are well documented for the various applications, including gas storage and separation, heterogeneous catalysis and proton conduction [39-43]. By incorporation of more advanced properties (e.g., luminescence, low framework density, open metal sites and functionalized pore for interaction), the research and development of these hetero-structured materials has been directed towards detection of pollutants [44,45], explosives and chemical warfare agents [46-48], multimodal imaging [49,50], and drug storage and delivery applications [51]. Luminescence characteristics can be introduced to MOFs either by metal ions or organic molecules (ligand), or through the incorporation of guest molecules [47,52-53]. In addition, the charge transfer from ligand to metal, metal to ligand and ligand to ligand as well as host-guest interaction between confined guest molecules and framework can also result in luminescence characteristics. The luminescence process and properties in MOF materials are discussed in detail in further sections.

9.1.1 Organic Nitro Compounds

ONCs are an important class of organic compounds derived from hydrocarbons by the replacement of one or more hydrogen atoms by nitro (-NO₂) groups. These organic nitro compounds are among the largest and most important groups of industrial chemicals in use today. The vast majority are synthetic, although several biologically produced nitroaromatic compounds have been recognized. The nitro group is strongly electronegative (electron-withdrawing) due to the combined action of the two electron-deficient oxygen atoms bonded to the partially positive nitrogen atom. When attached to a benzene or any other aromatic ring, the nitro group is able to delocalize π-electrons of the ring to satisfy its own charge deficiency. This not only provides charge to the molecule, but also imparts unique properties that make the nitro group an important functional group in various chemical syntheses. The nitro group is strongly deactivating toward electrophilic aromatic substitution, but facilitates nucleophilic aromatic substitution of the aromatic ring. Nitro group is often represented as containing a double bond between nitrogen and one oxygen and a co-ordinate or dative bond between nitrogen and other oxygen, however, in actual, the nitro group is stabilized by resonance and is a hybrid of two equal contributing structures, as shown in Scheme 9.1 [3,54].

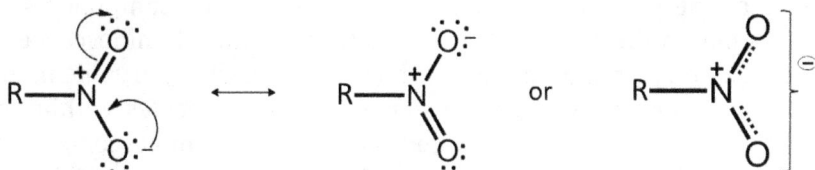

Scheme 9.1 Resonance in nitro group.

Generally, ONCs are very reactive molecules with high electron acceptor character due to electron deficient nitro group, thus, some of the ONCs show highly explosive nature such as RDX, 2,4-DNT, TNT and TNP (Figure 9.1). However, sensing of this class of explosives is not straight-forward because they have moderate vapor pressures and limited reactivity. Fortunately, these electron-deficient molecules are able to form π-stacking complex with electron-rich luminophores, and this particular property has been widely used for their recognition with fluorogenic probes such as conjugate polymers and luminescent metal organic framework (LMOFs).

9.2 Luminescent Metal Organic Frameworks

The use of fluorescent bridging ligands such as aromatic conjugated moiety, photoactive metal ions or combination of both allows the generation of luminescence properties in MOFs. Particularly, confinement of conjugated double bond ligands in the three dimensional solid state in combination with transition (d^{10}) or lanthanide metal ions generates luminescence in MOFs [55]. In these hybrid materials, the luminescence stimulates the emission of light by the π electron rich organic ligands often containing aromatic or conjugated moieties due to the absorption of the radiative excitation energy. In addition, d^{10}/ lanthanide metal cations are also responsible for luminescence in MOFs [56].

As depicted in Figure 9.2, the luminescence process in MOF materials is based on two types. Fluorescence, which is spin allowed transition from the singlet excited state S_1 to the ground state S_0 and typically characterized by a short-excited state lifetime in the order of 1-100 ns, or by a phosphorescence mechanism, in which intersystem crossing occurs from the singlet excited S_1 to the triplet excited state T_1, followed by a photon emitting forbidden transition to the ground state S_0, which is characterized by excited state lifetimes of 1 μs or longer. The confinement of inorganic and organic components in MOFs makes them a suitable candidate to produce luminescence in these solid crystalline materials. Metal ions/clusters, organic moieties, metal-organic charge transfer and guest molecules within porous/non-porous MOFs can potentially generate luminescence from a variety of mechanisms. For example, in ligand-centered emission, both photon absorption and emission processes occur on the same organic ligand. However, absorption and emission events also commonly occur in separate locations within the framework, with non-radiative energy and electron transfer processes explained by Förster–Dexter theory. Among these mechanisms, ligand-to-ligand charge transfer (LLCT), ligand-to-metal charge transfer (LMCT), metal-to-ligand charge transfer (MLCT) and metal-to-metal charge transfer (MMCT), guest centered emission and guest-sensitization mechanisms are the reasons behind the luminescence stimulation. These mechanisms are discussed in detail in the following sections.

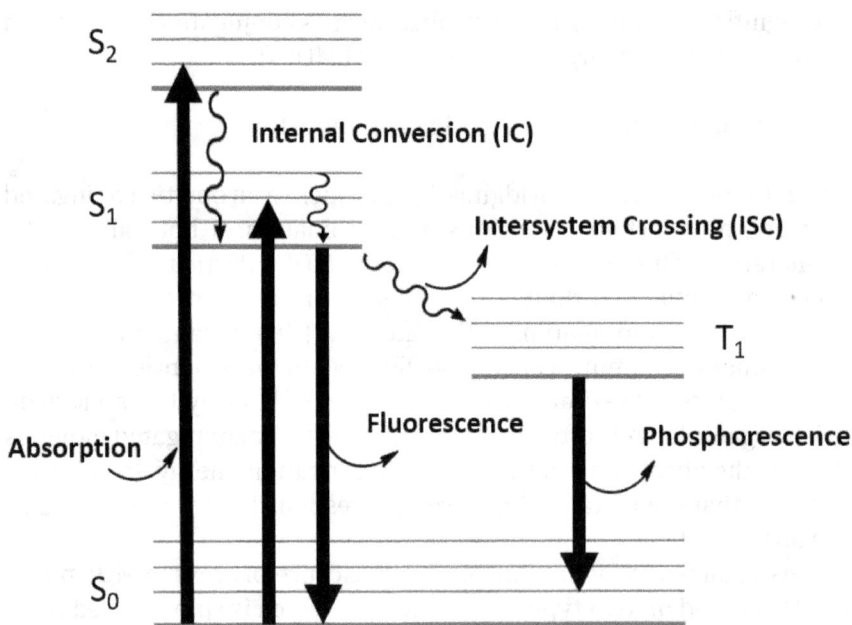

Figure 9.2 Jablonski diagram displaying schematic representations of various photophysical processes involved in the luminescence phenomena.

LMOFs have numerous key advantages over other potential luminescent probes. The inherent crystallinity of coordination polymers CPs/ MOFs allows for precise structural determination by X-ray diffraction, permitting exact knowledge of the atomic positions and interactions. Due to the advantage of porosity in many CPs/MOFs, LMOFs can easily allow the guest molecules (nitro analytes) into the pores that increases the interaction between the analytes and framework, thus, significantly improving the sensing ability in MOFs. The functionalization of pores or channels by using photochromic or conjugated organic ligands can increase the selectivity and sensitivity of MOFs towards analyte detection. Furthermore, post synthetic modification has also been demonstrated for adding functional groups to the framework to improve analyte detection [57,58]. Due to permanent porosity and cooperative luminescence properties of CPs/MOFs, LMOFs are particularly useful in the development of luminescent sensing materials for a variety of sensing applications.

9.2.1 Origin of Luminescence in LMOFs

Luminescence is the term used to explain the emission of light produced from the excited states upon the absorption of energy. The process of radiative relaxation from the excited states accompanying the emission of photons is categorized into two types of luminescence: fluorescence and phosphorescence. The light emitted between the energy states of the same spin multiplicity is called fluorescence (depicted in Joblonski diagram (Figure 9.2)). It is a spin-allowed transition occurring from a lowest singlet excited electronic state (S_1) to the ground state (S_0) with typical lifetimes in the order of nanoseconds (ns). Phosphorescence refers to the stimulation of light between states with difference spin multiplicity. In this case, the emission accompanying transition occurs from a lowest excited triplet electronic state (T_1) to the ground state (S_0). The inherent forbidden character of this electronic transition slows the emission process, and the corresponding lifetimes are found to be in the range of micro seconds to few seconds [56]. Metal organic frameworks basically comprise of two distinct kinds of potentially photoactive components: organic linker and metal cations or clusters. The occluded guest molecules in the porous cavities of certain MOFs are also responsible for luminescence. In most of the cases, organic molecules are the encapsulated guest species, whereas these are inorganic species in some instances. Thus, interpreting the observed luminescence signal in such systems demands consideration of the whole set of photophysical transitions. Generally, luminescence in LMOFs arises from the building blocks such as conjugated organic ligands or metal cations/clusters, as mentioned above. Organic linkers with aromatic moieties or extended π systems are commonly used in the design of porous MOFs due to their rigid molecular backbone. The π electrons in these linkers contribute greatly to luminescence by inter-stacking between adjacent linkers which can be classified as linker-based luminescence or LLCT [59]. Confined organic luminophores with an ordered arrangement in MOFs lead to close proximity, and the nature of their intermolecular interaction can be altered resulting in photoemissions that are different from their free form. Moreover, the emission of individual organic molecules is also affected in MOFs upon strong interaction with foreign guest molecules. The presence of secondary functional sites in MOFs, generally attached covalently to organic ligands, has been found to aid in expediting binding of guest molecules by weak but considerable non-covalent interactions. The confinement of such guest molecules has been found to have a remarkable effect in changing the emission of the ligand [52,53]. Another photoactive building unit in metal organic frameworks is metal ion/cluster. The luminescence property is largely dependent on the type of metal ions. Luminescence in MOFs comprising of transition metal ions in the framework is typically centered on the ligand rather than on the metal, but can also involve charge transfer between the metal and ligand. Emission in paramagnetic transition metal (unpaired electrons) based MOFs is not so strong because ligand-field transitions (d-d transition) may lead to strong re-absorption and/or quenching of luminescence, which can occur *via* electron or energy transfer through the partially filled d-orbitals. Exceptionally, some paramagnetic metal-based MOFs illuminate with strong emission under blue light (UV range), for example, Cu^{2+} based MOF comprising of pentiptycene based π conjugated ligands [60]. However, metals without unpaired electrons, especially d^{10} (Zn^{2+}/Cd^{2+} and Cu^+/Ag^+) metal-based MOFs, can produce strong ligand centered emission. Based on the fully filled d-orbitals of the transition metals, two types of charge transfer exist between the metal ion and organic ligand in MOFs: ligand to metal charge transfer and metal to ligand charge transfer. LMCT is often observed in Zn^{2+} and Cd^{2+} complexes [61,62], while MLCT is commonly seen in Cu^+ and Ag^+ compounds [63,64]. Metal-centered luminescence is often found in MOFs when lanthanide metal ions are incorporated in the rigid frameworks. This phenomenon is observed even in actinide-based MOFs [65,66]. Strong photon absorbing ligands with efficient intersystem crossing (ISC) are favored for constructing lanthanide LMOFs because they ensure the delivery of excitation energy through an antenna effect from their triplet excited states to the emissive states of lanthanides [56].

The various effects contributing to the luminescence of MOFs have been schematically demonstrated in Figure 9.3. A wide range of organic linkers has been utilized to stimulate luminescence in MOFs,

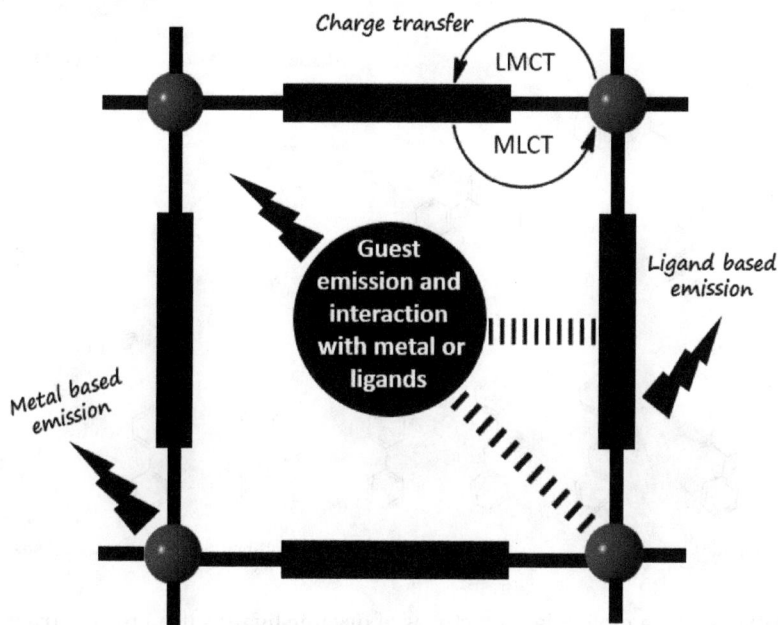

Figure 9.3 Representation of emission possibilities in the metal organic framework.

many of them have rigid backbones with di- or multi carboxylates [60,67]. Apart from polycarboxylates, several imidazole and pyridyl containing MOFs are also useful as the luminescent metal organic frameworks for various applications, including detection of organic nitro compounds [68,69]. The molecular structures of common linkers are listed in Figure 9.4 with abbreviations.

9.2.2 Advantages of Using LMOFs as Sensory Materials

Generally, LMOFs are often equated with the conjugate polymers due

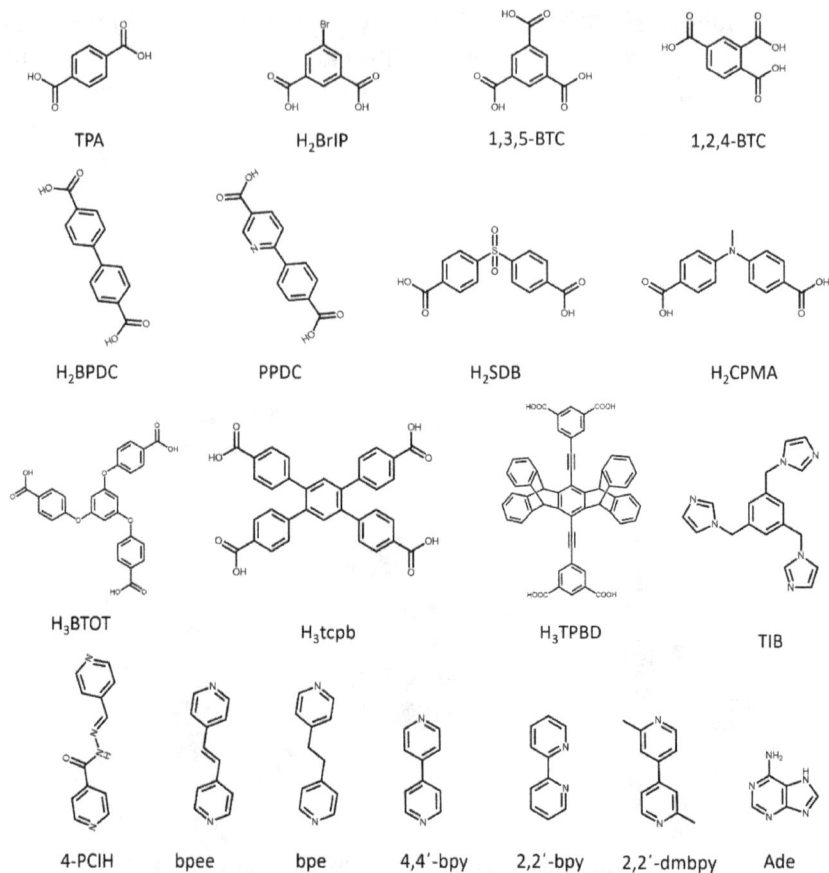

Figure 9.4 The molecular structures of organic ligands used to construct various types of LMOFs.

to their performance as sensory materials. When compared to the conjugate polymers, LMOFs exhibit significant advantages due to their crystalline nature, diverse and easily modifiable structures and topology, permanent porosity, wide range of physicochemical and electronic properties, etc. [70]. Permanent porosity with suitable functionalities are the most important characteristics in the field of molecular recognition. For any sensory material, sensitivity, selectivity, response time, stability, reusability and portability are the major parameters. The intrinsic porosity of MOFs provides large colonies for specific guest molecules which makes them suitable scaffolds at the molecular level. Interior pore surface of MOFs is more effective for the selective and sensitive recognition of specific molecules due to its molecular sieving effect. Functional groups within the framework, such as Lewis acidic or basic sites of ligands and/or open metal sites or functional groups which facilitate the hydrogen bonding interactions, further promote the desired analyte binding for selective detection. The encapsulation of the guest molecules in the pores not only increases the host-guest interactions, but also increases the selectivity and sensitivity of a specific sensor towards a particular analyte [71]. The response time is another primary characteristic of sensory materials, as the size, shape and nature of the pore surface in MOFs are directly linked to adsorption kinetics. Though MOFs exhibit better performance over other materials with respect to most of these parameters, however, the stability concerns, especially chemical inertness and hydrolytic stability, continue to prevent the development of practical implementation as most of the MOFs are hydrophilic in nature. Several approaches are currently being experimented to achieve moisture tolerance. Fluorescent conjugated polymers show diminishing emission intensity at elevated temperatures, whereas MOFs show strong emission even at relatively high temperatures. Thus, it is possible to utilize their fluorescence when the binding of a specific analyte requires an elevated temperature. Moreover, the electronic properties of the MOF structure may be fine-tuned. For example, given the same metal center and network connectivity, band gaps can be varied by changing the size of the SBU and the degree of conjugation of the organic linkers [72]. In fact, the changes in the framework's metal centers or their connectivity also varies the band gaps and atomic compositions of the valence band (or HOMO) and/or conduction band (or LUMO). Such tunability is vital for sensing applications, as it directly relates to the optical absorption and emissions properties. Furthermore, immobilization of organic ligands in a rigid framework can potentially reduce non-radiative relaxation triggered by free rotation and vibration of the linker, therefore, leading to stronger emissions [73-75]. Aggregation induced emission (AIE) is the best example, a low-emissive ligand in a dilute solution may show strong fluorescence emission upon confinement into a rigid frameworks of MOFs [61]. Finally, compared to amorphous materials, the highly ordered crystalline nature of porous MOFs allows precise identification and characterization of their structure by using X-ray diffraction, thus, making them suitable structures for investigating structure-property correlations and host-guest interactions. These significant attributes make these materials useful in both fundamental studies as well as practical sensing applications

9.2.3 Approaches and Strategies for Synthesis and Design of LMOF sensors

Rational design, control and construction are the major requirements for exploring and developing MOFs/LMOFs as sensory materials. MOFs are typically constructed by connecting SBUs with organic ligands, leading to various network topology structures. The organic spacers or the metallic SBUs can be altered to regulate the pore environment of the MOF material, which, in turn, may control its interactions with the encapsulated guest molecules and ultimately enables its utilization for a particular application. Since the MOF structure depends mainly on the choice of metals ions and ligands, various metals (transition, lanthanide/actinide and alkaline earth metals) with different size, coordination geometry and functional ligands have been utilized to tune the topology and architecture [76]. Generally, MOFs/CPs crystalline materials have been produced under solvothermal conditions *via* conventional electrical heating. Most MOFs syntheses are liquid-phase syntheses, where separate metal salt and ligand solutions are mixed together or a solvent is added to a mixture of solid salt and organic ligand in a reaction vessel. The yielded crystalline materials are influenced by various factors such as solvent, pH, temperature and molar ratio of reactants [77]. Apart from these liquid-phase synthesis techniques, solid-phase synthesis (mechanochemical) has also been documented as it is a quicker, easier and eco-friendly approach [78,79]. However, solid-state synthesis faces difficulties in obtaining single crystals and determining product structure, which is otherwise not an issue in the case of solution phase reaction methods. The slow evaporation/diffusion method is a regular process of crystallization which has been applied for the last few decades to prepare MOF crystals. Although routine synthesis of MOFs involves solvothermal methods, other methods such as microwave-assisted synthesis, mechanochemical synthesis, electrochemical synthesis and sonochemical synthesis have been applied as alternatives for MOF as well as LMOF synthesis [80-83]. These versatile synthetic techniques have facilitated the faster and effective production of MOFs/LMOFs for various applications.

MOFs have been widely explored for gas and hydrocarbon adsorption. Recently, significant advances have been made in the design of MOF based sensors for gas, VOS, toxic metal ions, nitro explosives and other analytes, exhibiting superb performance. When screening existing LMOFs as potential sensors, the ones able to selectively adsorb particular targeted analyte should be given priority. Accuracy for the adsorption of an analyte molecule is usually achieved through the precise construction of desired pores or surfaces at the molecular level [84,85]. The size and functionalization of pores or surfaces are most desirable as the only molecules smaller than the pores can be captured. Thus, controlling the pore size and functional groups on the surface is the most obvious step to consider when designing LMOF based sensors. The porous environment of the MOFs can be changed by various chemical manipulations to suit the guest molecules. Many physico-chemical properties within the porous structure, such as hydrophobicity, polarizability, polarity, acidity and proton affinity, can be finely tuned. By controlling the chemical environment of the pore, sensitive and selective capture of the targeted analytes can be achieved. Incorporation of Lewis basic sites in open frameworks facilitates the attraction of metal cations as well as acidic sites of picric acid such as -OH. In some particular MOFs with $-NH_2$ and $-CO-NH$, functional groups can facilitate hydrogen bonding interaction between the framework and guest analytes [86,87]. Open metal sites (OMS) are also attractive for sensing small molecules [88]. For example, the effective detection of NH_3 is realized by the preferential binding of this guest molecule at an OMS [89], and, in another case, terminal solvent molecules are assisted in sensing of anions through hydrogen bonding interactions between the solvent molecules and anions [90].

The electronic properties of LMOFs are also vital with respect to their sensing performance. Electron and/or energy transfer between a LMOF and an analyte are the main causes for a fluorescence response, and, as such, rational design of a LMOF should aim at promoting these features. The introduction of highly conjugated linkers to the skeleton of LMOF is expected to effectively attract aromatic or conjugated analytes through π–π interactions [91]. The relative orbital energies of the CB (or LUMO) of LMOFs can be altered by incorporating electron-donating or electron withdrawing groups into the ligands with different functionalities [92]. The energy transfer between the host and guest has proved that the overlap between the emission spectrum of LMOF and excitation or absorbance spectrum of a particular analyte induces the dramatic reduction in the fluorescence intensity of LMOF [93].

Thus, the target analyte detection is mainly depend on LMOFs characteristic such as pore size, pore/surface environment and electronic properties.

9.3 Detection of Organic Nitro Compounds by LMOFs

LMOF based sensory materials have been exploited for a variety of applications due to their unique ability to selectively detect and capture the targeted analyte molecules [63,93]. The permanent porosity of these robust insoluble solid materials makes them good scaffolds for analyte molecule capture by reversible adsorption and release [94]. Porosity of LMOFs not only encapsulates guest analyte molecules, but also provides an electron rich pore surface to accelerate the host-guest interaction [63]. Therefore, LMOF based detectors can be highly sensitive, recyclable and economically effective. LMOFs have been successfully used for the detection of volatile organic compounds (VOCs), small molecules and ionic species [46]. Owing to the high conjugated π-electron ligands in the framework, these materials have been widely studied for the detection of electron deficient explosives or explosives like molecules (ONCs). Exploiting LMOFs as sensors, a variety of nitrated organic compounds have been studied in vapor phase and in different liquid mediums [47]. Due to the nitro ($-NO_2$) functional groups, ONCs are electron deficient in nature and observed quenching in fluorescence intensity due to the interaction with luminescent MOFs is measured as a tool for the sensing of the analytes. The quenching efficiency for a particular analyte is calculated using $(I_0-I)/I_0 \times 100\%$, and the quantitative calculation of the quenching efficiency of a particular LMOF sensor for a targeted analyte in liquid medium can be evaluated by using Stern-Volmer analysis [86,95] as

$$I_0/I = K_{SV}[Q] + 1 \text{ (Linear)}$$

$$I_0/I = Ae^{k[Q]} + B \text{ (Non-linear)}$$

In above equations, I_0 and I are the fluorescence intensities before and after the addition of the respective analyte, $[Q]$ is the molar concentration of the analyte and K_{SV} is the quenching constant or Stern-Volmer constant of a particular LMOF. The higher K_{SV} value indicates a high efficiency of the sensor. For the detection of organic nitro compounds, a variety of LMOFs based on different transition metal ions, lanthanide metal ions and main group metal ions in combination with aromatic organic ligands have been successfully explored. In contrast to other metal ions, transition/lanthanide metal ions have been utilized to construct LMOFs to a greater extent. Among transition metal ions, Zn^{2+} and Cd^{2+} are commonly employed in addition to Cu^{+}/Ag^{+} because the d^{10} metal ions not only show different coordination and geometries, but also exhibit luminescence properties when bound to ligands. Li *et al.* [5] first reported the detection of organic nitro compounds by Zn based LMOF in 2009.

9.3.1 Detection by Zn based LMOFs

Due to the d^{10} metal ions, Zn based MOFs comprising of fluorogenic ligands with π-conjugated moieties are good candidates for designing luminescent MOFs. The origin of luminescence in Zn-LMOFs can be attained either by a single ligand or *via* a ligand-to-ligand charge transfer process. Confinement of ligands in robust 3D frameworks often enhances the ligand interactions and alters the HOMO-LUMO energy of the resultant structures. This leads to different emission behaviors with respect to the molecular state which can alter the luminescence intensity of the free form of the ligands [46,56,96]. Li and coworkers reported the very first Zn based luminescent chemical sensor for explosive detection in 2009 [5]. $[Zn_2(bpdc)_2(bpee)]\cdot2DMF$ (LMOF1) is a 3D interpenetrated microporous LMOF with permanent and sustainable one-dimensional (1D) channels. Crystal structure revealed that the three-dimensional framework's 1D open rectangular channels are along the crystallographic [010] direction with encapsulated DMF molecules. The outgassed framework shows permanent porosity (pore size 7.5 Å and pore volume 0.17 cm³/g) with Langmuir surface area of 483 m²g⁻¹. The activated sample shows strong solid-state fluorescence at 420 nm (λ_{ex} = 320 nm) at room temperature. The vapors of two representative aromatic and nitroaliphatic analytes, DNT and DMNB, exposed to the thin layer of solid samples on glass slides showed rapid and reversible response and the quenching percentage reached maximum (~85 %) within 10 seconds for both DNT and DMNB analytes. The exceptional sensitivity and fast response of the framework towards the two analytes can be attributed to its 1D open channels which facilitate fast diffusion of the explosive molecules on the internal pore surface. Also, its extended 3D framework allows higher number of analyte binding sites in the channels which enhances quenching for both nitro analytes. In addition, the particle size effect has also been studied; smaller particles (layer thickness ~5 μ) can decrease the quenching time by 24 times with respect to larger particles (layer thickness ~30 μ). Further, the authors have studied the detection of NACs explicitly in vapor phase with another zinc based LMOF, [Zn₂(oba)₂(4,4′-bpy)]·DMA (LMOF-2) [6]. The 3D porous structure was built on $Zn_2(oba)_4$ paddle-wheel SBU (Figure 9.5a-b). As depicted in Figure 9.5c, each SBU is connected by four oba ligands to form a 2D distorted 4^4 net. The interpenetration of two such identical nets yields to the 2D layered network (Figure 9.5d). The 4,4′-bpy ligands coordinate to the paddle-wheel units of the adjacent layers, generating a robust 3D framework with two intersecting 1D channels along the crystallographic [100] and [010] axis, and the pore aperture of the channels is found to be ~5.8 × 8.3 Å.

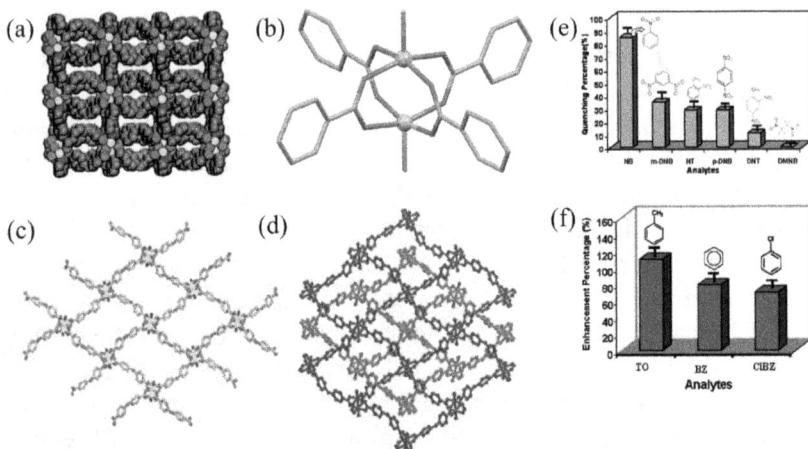

Figure 9.5 Crystal structure of [Zn₂(oba)₂(4,4′-bpy)]·DMA. (a) Space-filling model of the 3D framework, illustrating the 1D channels running along the [100] axis, (b) SBU [Zn₂(CO₂)₄] connected by four oba ligands, (c) a distorted 4^4 net of Zn(oba) layer, (d) two-fold interpenetration, shown by two different colors (blue and red) and (e-f) percentage of fluorescence quenching by electron withdrawing groups; percentage of fluorescence enhancement of electron donating groups. Reproduced from Reference 6 with permission from American Chemical Society.

The outgassed sample stimulates strong emission at 420 nm upon irradiation at 280 nm. The compound shows opposite behavior when exposed to aromatic molecules with different electronic properties, electron withdrawing (nitro functional) and electron donating groups. Among different nitro analytes (NB, NT, 1,3-DNB, 1,4-DNB, 2,4-DNT and DMNB), NB shows highest quenching ability because of its high equilibrium vapor pressure and electron withdrawing –NO₂ group, whereas NT quenching performance is much lower (29%) as compared to NB because of the presence of electron donating group

(-CH$_3$), even though the vapor pressure of NT is comparable with NB. On the other hand, all aromatic analytes with electron donating group show enhancement in luminescence intensity, and the extent of such enhancement is consistent with their electron donating ability and vapor pressure (Figure 9.5e-f). In both cases, the emission intensity of the host framework can be fully regenerated by heating the exposed sample at 150 °C for a few minutes.

Ghosh *et al.* [91] synthesized a 2D micro porous luminescent metal organic framework, [Zn$_{1.5}$(BTOT)(H$_2$O)] (LMOF3), with a starfish like electron rich flexible tripodal ligand (BTOT). Supramolecular wire effect was utilized to improve detection of both aliphatic (DMNB) and aromatic (DNT) nitro analytes in the vapor phase. SXRD studies revealed that two-dimensional networks were well arranged in 3D lattice with strong π-stacked cylindrical units of orderly arranged BTOT moiety. LMOF3 consists of bilayer sheets of the tripodal ligand connected by linear trimeric Zn^{2+} in a bidentate fashion. The 2D sheets contain hexagonal pores with approximate dimensions of 12.2 × 13.4 Å^2. Trimeric Zn^{2+} units of alternate sheets are positioned at the center of the hexagonal pores of the middle sheet from above and below, forming a starfish array of the framework. The 2D sheets of LMOF are arranged in an A-B-A-B-A fashion. The π-π stacked 1D molecular packing of the ligands, leading to a supramolecular wire like arrangement, enhances luminescence properties in the 2D luminescent materials. A thin layer of luminescent material showed strong solid-state fluorescence at 390 nm upon excitation at 280 nm at room temperature. Exposure of the thin layer to nitro aromatic and aliphatic compounds, NM, NB, NT, DMNB, 2,4-DNT, 2,6-DNT and 4-NP for 10 min, showed different degrees of emission quenching. Due to strong π-π interactions with nitrobenzene, the emission intensity was almost quenched by NB (86%), followed by NM (75%). The overall order of the quenching efficiency with all nitro analytes was found as followed, NB > NM > NT > DMNB > 2,4-DNT > 2,6-DNT > 4-NP, with their respective quenching percentages 86, 75, 66, 57, 53, 46 and 39. The observed trend can be explained based on several factors such as the electron withdrawing ability, electronic properties and vapor pressure of the analytes, along with the supramolecular wire effect imposed by the host framework. The enhanced quenching efficiency of the nitroaliphatic analyte DMNB, which lacks any conjugated π system, is described by the orbital overlap between the electron rich π system of the host LMOF and the electron deficient −NO$_2$ groups of the analyte. The long range migration of excitons by supramolecular wire effect facilitated such an overlap, in addition to the strong electronic and C-H···O interactions with the aliphatic nitro analytes. Due to smaller size of nitro methane, the observed quenching might be influenced by steric effect, as NM can access the interlayer space readily over bulkier analytes such as 4-NP. LMOF3 showed high efficiency for both nitro explosives taggants (2,4-DNT and DMNB) within a short period of time.

Chen and coworkers reported sensing capacity of $[Zn_4(OH)_2(1,2,4-BTC)_2(H_2O)_2]\cdot0.63DEF\cdot3.5H_2O$ (LMOF4) in selected organic solvents including NB in solution phase [97]. The 3D framework contains 1D open channels of a cross section of ~6.2 × 5.7 $Å^2$ along the a-axis. The free and coordinated solvent molecules occluded in the channels can be removed by soaking in fresh solvent, followed by thermal activation. The permanent porosity of the desolvated luminescent material is confirmed by gas adsorption experiment, which exhibited a BET surface area and pore volume of 408 m^2g^{-1} and 0.205 cm^3g^{-1}, respectively. The simulated pattern of the crystal structure is in well agreement with the experimental patterns of the activated material, and the framework remains stable upon activation, as evident from the PXRD analysis. The methanolic solution of the activated sample shows strong quenching effect to NB vapors. The encapsulation of the analytes into the porous structure is confirmed by the IR spectra of the samples after nitrobenzene soaking. The quenching behavior of luminescent material with NB molecules can be attributed to the charge transfer effect from the aromatic linkers of the LMOF to the electron deficient guest molecules as well as π–π interaction between the host framework and analyte molecules.

In another study, Cui *et al.* [98] studied the performance of a metal–metallosalen based microporous LMOF for the detection of volatile nitro organic vapors. The solvothermal reaction of dicarboxyl-functionalized salen ligand pdbs and 4,4′-bpy resulted in $[Zn_2O(Znpdbs)_2(bpy)(DMA)(H_2O)]\cdot4DMA\cdot4H_2O$ (LMOF5), a double interpenetrated 3D microporous MOF. The secondary building unit, $Zn_2(\mu_2-O)(CO_2)_4$, bridged by four Zn-pdbs units and 4,4′-bpy pillar linkers formed a 3D framework of a primitive cubic net with large parallelogram channel having diagonal distances of ~23.0 × 31.2 Å along the b-axis. The rectangular open channels were activated, and the permanent porosity was measured by N_2 adsorption which exhibited only surface adsorption with a BET surface area of 10 m^2g^{-1}. The thin-layer of the activated sample exhibited a strong fluorescence emission at 535 nm upon excitation at 420 nm. Among the vapors of different nitro analytes (NB, 2-NT and 3-NT) subjected to the thin layer of the sample, NB exhibited the strongest quenching effect due to its strongly electron withdrawing -NO$_2$ group and the highest vapor pressure. Within 10 seconds, NB quenched the emission by 67% and the quenching percentage reached 75% on prolonged exposure up to 4 min. Additional studies on a series of LMOFs (Zn$_3$(bpdc)$_3$(4,4'-bpy)·4DMF·H$_2$O) (LMOF6), Zn$_3$(bpdc)$_3$(2,2'-dmbpy)·4DMF·H2O (LMOF7), Zn$_2$(bpdc)$_2$(bpe)·2DMF (LMOF8) and Zn(bpdc)(bpe)·DMF (LMOF9)) having strong fluorescence emissions were also carried out. This series of LMOFs was generated using similar linker ligands, however, had different framework topology. Further, PL responses of LMOFs were monitored towards electron deficient and electron rich analytes. The results showed that the materials spontaneously responded to electron deficient analytes through fluorescence quenching and to electron rich analytes through fluorescence enhancement, respectively [99].

Coordination polymers based on conjugated angular dicarboxylate ligands in combination with d^{10} metal ions like Zn^{2+} are expected to display good photoluminescence properties both in the solid and suspension state. Suresh *et al.* [100] synthesized a luminescent MOF material {[Zn$_3$(μ-OH)$_2$(SDB)$_2$]·(PPZ)}$_n$ (LMOF10) and investigated its sensing behavior with nitro organic molecules. LMOF10 was constructed by paddle wheel type trinuclear clusters of divalent zinc and SDB ligands. Crystallographically unique Zn atoms are linked by four carboxylate groups in *syn-anti* fashion from different SDB ligands to generate a two-dimensional double loop coordination polymeric network (Figure 9.6a-b). LMOF10 was activated in order to evacuate guest molecules and gases, prior to detection of nitro phenols. The guest free framework retained its crystallinity and thermal stability with subtle changes in the PXRD peaks indicating structural changes upon expulsion of guest molecules from the framework. The activated samples were used to study the fluorescence properties of four nitro phenols, TNP, 2,4-DNP, 2-NP and 4-NP, in acetone medium. LMOF10 showed good sensing behavior with all nitro analytes due to their electron withdrawing nature and electron rich framework. The maximum fluorescence intensity of LMOF10 was reduced by 56% by TNP as compared to other nitro analytes 2,4-DNP, 2-NP and 4-NP, which showed poor fluorescence quenching of only 15%, 15% and 13% (Figure 9.6c-d). The quenching of the fluorescence intensity and a slight shift in the emission wavelength during the addition of TNP can be attributed to the supramolecular and electrostatic interactions between the nitro phenols and LMOF through interior and surface sites of the luminescent material.

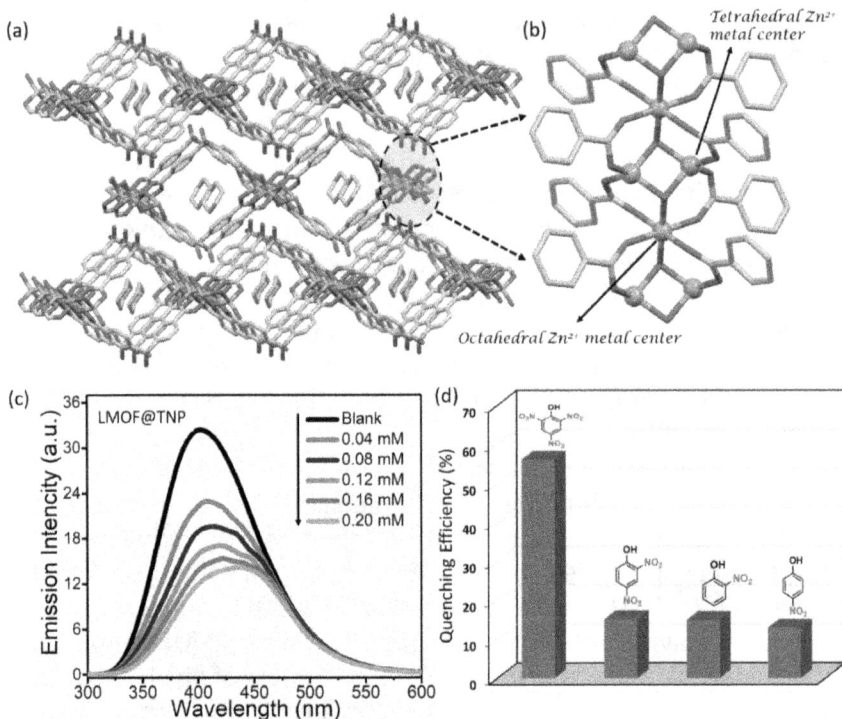

Figure 9.6 (a) Packing diagram of {[Zn$_3$(μ-OH)$_2$(SDB)$_2$]$_n$, depicting the staggering of adjacent 2D sheets and through cylindrical channels down the b-axis, (b) close-up view depicting the coordination environment around Zn^{2+}, (c) emission spectra of LMOF10 with incremental addition of 2 mM solution of TNP in acetone (λ_{em} 402 nm, λ_{ex} 290 nm) and (d) percentage of fluorescence quenching efficiency. Reproduced from Reference 100 with permission from Royal Society of Chemistry.

Confinement of electron rich π-conjugated ligands in 3D MOFs is useful to develop efficient luminescent materials in the field of sensory materials. Konar *et al.* [101] synthesized {[Zn(tcpb)$_{0.5}$(bpeb)$_{0.5}$][0.5(bpeb)·2H$_2$O]}$_n$ (LMOF11) based on highly conjugated rigid bpeb linker in combination with a π-electron rich tetratopic carboxylate ligand H$_4$tcpb for the detection of nitro organic compounds. The crystal structure of LMOF11 reveals that the crystallographycally unique Zn^{2+} center adopts a square pyramidal geometry with four oxygen atoms from four different tcpb ligands and one nitrogen atom from the bpeb ligand. The structure of LMOF11 is constructed by a paddle wheel type SBU ([Zn$_2$(tcpb)$_4$(bpeb)$_2$]) with two Zn^{2+} centers, four tcpb and two bpeb ligands. Each tcpb ligand is connected to four SBUs and forms a 2D layer on the *bc*-plane which is further pillared by bpeb, resulting in a 3D interpenetrated framework with 1D channels along the *b*-axis. In this 3D structure, two types of bpeb ligands has been found, one coordinated to Zn^{2+} metal ion and other remains uncoordinated in the 1D channels. The ligand arrangement as well as the π-electron rich environment in the entire framework facilitated the use of LMOF11 towards the sensing of nitroaromatic explosives. The photoluminescence spectrum of the material dispersed in DMF exhibits a strong fluorescence band at 415 nm with a small shoulder at 393 nm upon excitation at 340 nm, which is attributed to the LLCT between the free and coordinated bpeb. The luminescent properties were further examined by dispersing LMOF11 in various nitro explosive molecules such as NB, 4-NT, 1,2-DNB, 1,3-DNB, 1,4-DNB, 2,4-DNT, 2,6-DNT and PA. The fluorescence intensity was highly dependent on the concentration of PA, with an effective sensing behavior and quenching of the initial fluorescence intensity by 94%. On the other hand, NB, 2,6-DNT and 2,4-DNT had a minor effect on the emission intensity of LMOF11, whereas 4-NT, 1,2-DNB, 1,3-DNB and 1,4-DNB showed a negligible effect. From the linear fitting of S-V plot, the quenching constant of LMOF for PA was found to be 8.1×10^4 M^{-1}, which signaled the superior quenching ability towards PA.

Besides detection of nitro explosives in organic phases, Zn based LMOFs have also been investigated in aqueous phase. Our group reported a mixed ligand {[Zn(TPA)(4-PCIH)]·3H$_2$O}$_n$ (LMOF12), which is constructed by 4-PCIH and TPA ligands [102]. The dimeric [Zn$_2$(COO)$_2$] secondary building unit formed by carboxylate oxygen atoms of 1,4-BDC (TPA) in μ$_2$-coordination mode create square lattice (*sql*) sheets which are further connected by pyridyl nitrogen atoms of axial ligand with overall network topology exhibiting distorted uninodal *pcu* (Figure 9.7a). The water stable luminescent material exhibited a strong emission at 458 nm upon irradiation at 285 nm at room temperature. The luminescent properties have been investigated with different pools of nitro analytes, TNP, 2,4-DNP, 4-NP, 2-NP, 2,4-DNT, 2,6-DNT, 4-NT, 3-NT, 2-NT, NB, NM and DMNB, in aqueous media. Upon incremental addition of 0.04, 0.08, 0.12, 0.16 and 0.20 mM TNP to the aqueous solution of LMOF12 (Figure 9.7c-d), the observed fluorescence quenching efficiencies were 43, 55, 60, 65 and

70%. Noticeably, equivalent amounts of other analytes resulted in relatively low or negligible quenching efficiencies. The order of the quenching efficiency was as follows, TNP > 2,4-DNP > NP, which is in agreement with the order of their acidity. TNP showed fast and selective fluorescence quenching as high as up to 70% (Figure 9.7e).

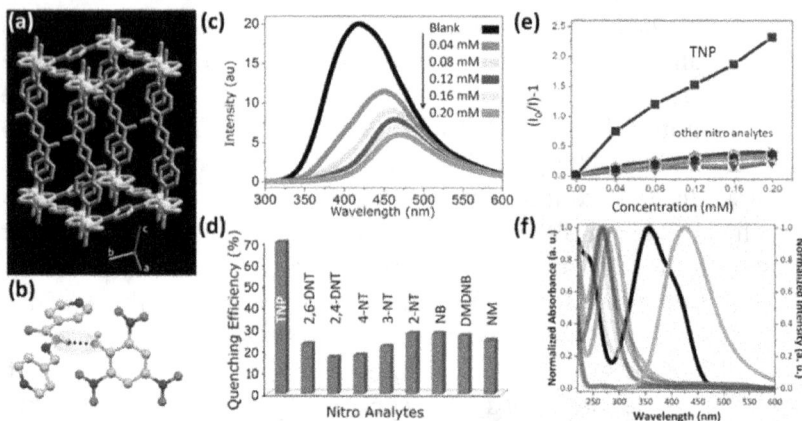

Figure 9.7 (a) Schematic depiction of metal-organic cube imparting distorted pcu topology to the framework of {[Zn(TPA)(4-PCIH)]·3H$_2$O}$_n$, (b) crystal structure depicting the strong N-H···O hydrogen bonding interaction between TNP and 4-PCIH ligand, (c-d) fluorescence spectra at different concentrations of TNP and S-V plot of various nitro analytes, (e) quenching percentages of different nitro analytes and (f) spectral overlap between normalized absorbance spectra of nitro analytes and normalized PL spectra of corresponding LMOF. Reproduced from Reference 102 with permission from Wiley-VCH

At lower concentrations, LMOF12 can detect TNP up to 82 ppb (0.36 mM), and the ability of quenching was calculated to be 1.16×10^4 M^{-1}. According to linear S-V plots of LMOF with all nitro analytes including TNP, the quenching mechanism is predominantly static in nature which is also further supported by the fluorescence decay lifetime experiments. The average excited-state lifetime (τ) values before and after addition of TNP are 0.72 ns and 0.71 ns. The fluorescence decay lifetimes do not change in the presence of TNP, indicating that the static quenching mechanism. In addition, the spectral overlap of the emission and absorbance spectra suggests that the resonance energy transfer mechanism is also operative in the quenching process (Figure 9.7f). The energy transfer for the quenching mechanism is also attributed by the N-H···O as well as C-H···O hydrogen bonding interactions between the phenolic group of TNP and carbonyl hydra-zone moiety of axillary ligand (Figure 9.7b).

Recently, we developed Zn-adenine based bio-luminescent poly-mer, [Zn(μ_2-1Hade)(μ_2-SO$_4$)] (LMOF13), which showed selective and sensitive detection of TNP over the large pool of nitro analytes in aqueous medium [103]. LMOF13 is 2D coordination polymer com-prising of Zn^{2+}, μ_2-SO$_4$ and adenine molecule. SXRD data revealed that both adenine and sulfate units act as bridging ligands in the formation of the 2D network oriented in the *bc*-plane (Figure 9.8a). The 2D nets are stabilized in 3D lattice by intermolecular N–H$\cdots$O interactions with the adjacent sheets from either side (Figure 9.8b). The robust 2D framework showed good thermal stability up to ~450 °C and chemi-cal stability in various solvents including water. The solid-state pho-toluminescence spectra of LMOF13 showed strong emission at 416 nm upon excitation at 295 nm at room temperature. The aqueous sus-pension of LMOF13 also showed effective emission, almost similar to

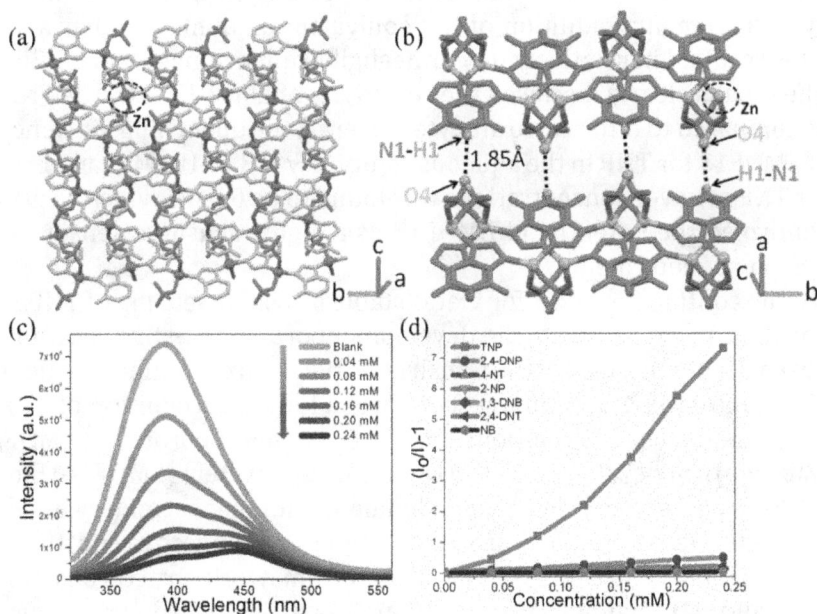

Figure 9.8 (a) Crystal structure of [Zn(μ_2-1H-ade)(μ_2-SO$_4$)]$_n$ oriented along the bc-plane, (b) observed intermolecular hydrogen bonding interactions (N–H$\cdots$O) between adjacent 2D sheets, (c) change in the fluorescence intensity upon incremental addition 2 mM TNP solution and (d) Stern-Volmer plots upon gradual addition of NACs. Reproduced from Reference 103 with permission from American Chemical Society.

solid state. Aqueous phase detection of nitro organic compounds was carried out using a variety of nitro analytes. The aqueous solutions of TNP, 2,4-DNP, 2,4-DNT, 1,3-DNB, 4-NT, 2-NP and NB were examined with the suspensions of LOMF13 [aqueous solutions of NACs (0.04 to 0.20 mM) with 2 mg/2 mL of water in a cuvette with constant stirring].

Due to electron withdrawing nature of all nitro analytes, every molecule acts as a fluorescence quencher, and the order of the quenching effect is in accordance with their electron withdrawing power. Among the different nitro analytes, TNP showed significant quenching behavior (88%), whereas it was 33% for 2,4-DNP. The quenching percentages were found to be in the following order: TNP > 2,4-DNP > 2,4-DNT > 4-NT > 1,3-DNB > 2-NP > NB and the respective quenching percentages were 88, 33, 21, 16, 11, 9 and 3%. Upon incremental addition of TNP solution from 0.04 to 0.24 mM to an aqueous suspension of LMOF13, the observed fluorescence quenching was 30, 55, 69, 79, 85 and 88%, respectively (Figure 9.8c). However, incremental addition of the equivalent amounts of other analytes resulted in relatively low or negligible quenching (Figure 9.8d). The results clearly disclosed the sensitivity of LMOF13 towards TNP, as compared to other nitro analytes. Overall, the quenching efficiency of LMOF13 for TNP in the aqueous solution was 3.1×10^4 M^{-1}, and LOD for TNP at lower concentrations was found to be 0.4 nM which clearly confirmed the potential of LMOF13 as a highly sensitive sensor for TNP in aqueous media.

The continuous quest for water stable MOFs for sensing of TNP in aqueous media has led to the development of an interesting functionalized 3D luminescent MOF. Ghosh *et al.* [87] utilized amine functionalized Bio-MOF-1 as a luminescent probe for the recognition of TNP in aqueous solutions. The microporous MOF, $[Zn_8(ade)_4(bpdc)_6O \cdot 2Me_2NH_2] \cdot G$ (G = DMF and water) (LMOF14/Bio-MOF-1), possessed infinite zinc-adeninate columnar secondary building units, interconnected through multiple bpdc linkers to constitute a 3D extended framework. The overall framework structure is anionic and 1D channels present along *c*-axis are filled with dimethylammonium $[Me_2N(+)H_2]$ cations (the product of DMF decomposition) as well as DMF and water from the solvents used during synthesis. The disordered Me_2NH_2 cations can permit the facile diffusion of the analytes inside the porous framework, resulting in the close proximity of the electron-deficient analytes to the electron-rich MOF scaffolds. As established by several research groups, Bio-MOF-1 is highly stable in aqueous solutions and exhibits structural stability in aqueous TNP solutions, as also confirmed by PXRD analysis. The desolvated material excited at 340 nm and exhibited a strong emission at 405 nm. Benefiting from the stability in water, Bio-MOF-1 was examined for the sensing of the trace amounts of nitro explosive molecules in aqueous medium. The emission response was obtained by fluorescence titration with different nitro explosives, such as TNP, TNT, RDX, DMNB, 2,4-DNT, 2,6-DNT and NM. The results showed sensitive and rapid detection and high fluorescence quenching (93%) for TNP on incremental addition of 10^{-3} M (200 µL) analyte. This clearly demonstrates that Bio-MOF-1 exhibited selectivity towards TNP over other nitro analytes. It could also detect TNP at extremely low concentrations, with a detection limit of 12.9 nm (2.9 ppb). As discussed in the previous example about the quenching mechanism, the interaction between the framework and TNP has been clearly evidenced by the co-crystal of TNP and functional coligand (adenine). The free primary amine group of adenine forms a cooperative intermolecular hydrogen-bonded complex with TNP with strong N-H···O interaction (Figure 9.9). In addition, dimethylammonium cations can also enhance the intermolecular interactions with the guest nitro analytes. The complexation process and smooth diffusion in 1D channels can magnify the host-guest interactions, giving rise to a highly sensitive response to TNP over other nitro analytes.

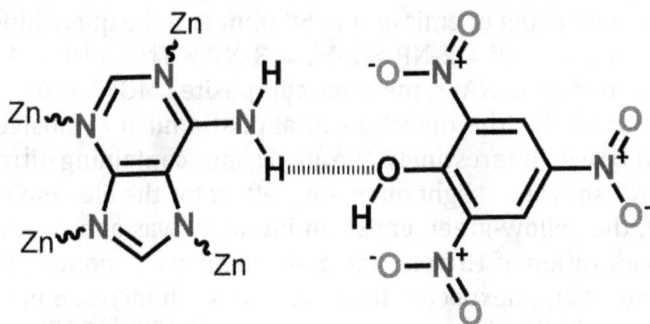

Figure 9.9 Plausible H-bonding interaction between adenine of Bio-MOF-1 and TNP.

However, compared to nitro aromatic compounds, nitro aliphatic compounds have lower saturated vapor pressure and undergo much weaker interactions with the host material. Thus, developing a luminescent material for detecting aliphatic nitro explosives, including highly explosive nitro compounds like, RDX, HMX and DMNB, is a challenge. Recently, Zang and coworkers developed a green fluorescent LMOF for the detection of aromatic and aliphatic highly explosive nitro compounds [104]. The luminescent MOF was the composite of anionic dye 8-hydroxy-1,3,6-pyrenetrisulfonicacid (HPTS) and $[Zn(TIPA)(NO_3^-)_2] \cdot 5H_2O$ (LMOF15) cationic metal organic framework. The porous cationic framework was synthesized from the self-assembly of Zn^{2+} ions and neutral TIPA along with NO_3^- as counter ions. The anionic dye HPTS ions were introduced into the channels of MOF *via* an ion-exchange approach. The composite material showed a strong dual emission fluorescence benefiting from the confined pores and dispersion effect. The peak emission of the composite LMOF was close to that of TIPA, which displayed an intense and broad emission between 375 and 450 nm. The solid state photoluminescence spectra of the composite LMOF showed a dual-emitting behavior at around 375-450 and 500-575 nm. The water stable composite LMOF was studied for the sensing properties of several nitroaromatic and nitroaliphatic explosives in the aqueous solution.

It has been observed that dual-emission peaks of the composite material decrease with increasing concentration of the nitro-explosives. Also, it is observed that the reduction of the fluorescence emission of HPTS is much more obvious than that of TIPA. 2-NP, 3-NP, 4-NP and TNP showed much better quenching effect than other nitro analytes. The concentrations of 2-NP, 3-NP, 4-NP and TNP caused the complete quenching of emission at 50 ppm and the quenching order was as follows: 4-NP > 2-NP > 2-NT > 3-NP > TNP > NB > 1,3-DNB. Besides detection of NACs, the same composite LMOF has been investigated further for the detection of aliphatic nitro explosives RDX, HMX and DMNB. Interestingly, the alkyl chain-containing nitro explosive DMNB showed a slight quenching effect for the blue emission, in contrast, the yellow-green emission intensity was almost quenched at a concentration of 15 ppm. Cycloalkane nitro compounds RDX and HMX showed an unexpected fluorescence. With increase in the concentrations of RDX, the emission intensity of HPTS at 525 nm diminished dramatically, but the fluorescence of the blue emission at 400 nm was greatly strengthened. At RDX concentration of 15 and 30 ppm, the yellow-green emission intensity was quenched by 31% and 55%, but the blue emission was enhanced by 40% and 195%, respectively. Moreover, RDX facilitated almost complete quenching of the emission of confined HPTS at a low concentration of 39 ppm. The obtained results clearly suggest that composite LMOF can act as a dual-emitting probe for the endorsed recognition of aliphatic nitro explosives in the aqueous phase.

9.3.2 Detection by Cd based LMOFs

Besides zinc containing LMOFs, cadmium based MOFs have also attracted intensive interest for potential applications in chemical sensors, photochemistry and electroluminescent display [93,105,106]. Several Cd based LMOFs have been studied for detection of nitro organic compounds. Cadmium and zinc have similar physical properties as they are the members of the same row in the periodic table. Due to toxicity related issues, cadmium based MOFs or luminescent sensors are limited to few applications.

Recently, an interesting cadmium based LMOF has been investigated by Sun and coworkers for the sensing of NB vapors [107]. $\{Cd_3(HCOP)(2,2'-bpy)_2 \cdot 4DMA\}_n$ (LMOF16) sensory material was synthesized by mixing a hexacarboxylate ligand HCOP with $Cd(NO_3)_2 \cdot 4H_2O$ and coligand $2,2'$-bpy *via* solvothermal method in mixed solvent DMA/H_2O at 100 °C for 3 days. The resulting crystalline material was structurally analyzed by SXRD and other physico-chemical techniques. SXRD analysis revealed interesting structural aspects: (i) the trinuclear Cd^{2+} secondary building unit of cadmium having two crystallographically different metal centers and (ii) Cd(1) is hepta-coordinated by two N atoms of $2,2'$-bpy ligand and five carboxylate oxygen atoms from three HCOP ligands in a distorted $\{CdN_2O_5\}$ pentagonal bipyramid geometry. The Cd(2) metal center adopted the typical $\{CdO_6\}$ octahedral coordination mode with six carboxylate oxygen atoms of *anti–anti* coordination conformations from four $HCOP^{6-}$ ligands. The trinuclear SBU is connected to four adjacent units *via* four $HCOP^{6-}$ ligands to form a 2D double layer, which stacked together and resulted in the overall 3D structure with 1D cylindrical open channels having approximates sizes of 16.3×11.7 $Å^2$ and 11.5×10.9 $Å^2$ along *c*-axis and 5.1×8.5 $Å^2$ along *b*-axis. The simplified network topology is (4,4)-connected uninodal *sql* double layer type network. The open 1D channels were occluded by DMA solvent molecules which were subsequently evacuated by soaking in methanol and DCM solvents for 3 days followed by heating at 80 °C for 15 h. The guest free framework exhibited a single step type-I N_2 adsorption curve with BET and Langmuir surface areas of 475.0 and 698.6 m^2g^{-1}, respectively and the total pore volume of 0.25 cm^3g^{-1}. Photoluminescence properties of the activated compound were investigated for the detection of NB vapors. The broad luminescence emission was centered at 427 nm upon excitation at 324 nm. The fluorescence emission was attributed by LLCT, admixing with MLCT for such Cd^{2+} coordination complexes. In order to detect vapors of NB, a thin layer of luminescent material on a quartz slide was exposed to NB, 2,4-DNT and NM as well as water and some organic solvents like methanol, ethanol, CH_2Cl_2, CCl_4, 2-propanol, phenylmethanol, benzene, toluene, bromobenzene, 1,2-dichlorobenzene, 4-chlorotoluene, etc., for 24 h. A strong turn-off fluorescence quenching was observed with NB (90% quenching percentage). Among the organic solvents, benzene showed nearly 8 fold enhancement in the emission intensity within 2-3 min. After detection of nitrobenzene and benzene, samples were recovered by simply heating at 80 °C under vacuum for 3 h.

Suresh and coworkers synthesized a cadmium-based coordination polymer and studied its luminescence properties with different nitro and non-nitro organic solvents in vapor phase [108]. $\{[Cd_3(1,3,5\text{-}BTC)_2(TIB)_2(H_2O)_4]\cdot(H_2O)_2\}_n$ (LMOF17) was constructed by benzene tricarboxylate (1,3,5-BTC) and flexible imidazolium based tripodal ligand (TIB). Owing its d^{10} metal node (Cd^{2+}), the activated material, on irradiation with different excitation wavelengths (330, 348, 374 nm), exhibited effective emission profile in the range from 425 to 460 nm. Excitation at corresponding higher energy absorbance wavelength (λ_{ex} 330 nm) showed weaker red-shifted emission bands in the range of 428-414 nm. However, strong red-shifted emission profile (λ_{em} 433 nm at λ_{ex} 348 and λ_{em} 446 nm at λ_{ex} 374nm) was observed at low energy excitation wavelength. The activated material showed different degrees of quenching in its photoluminescence intensity when exposed to small organic molecules such as methanol, toluene and nitro functionalized NM as well as NB. Considerable quenching was perceived when exposed to organic molecules containing nitro groups as compared to other analytes. The emission intensity was dropped by 69.3% and 58% for nitromethane and nitrobenzene, respectively. The selective sensing could be attributed to the narrow pore structure LMOF, which suitably facilitated the approach of less bulky analyte (NM) towards the metal centers compared to the bulkier analyte (NB).

Reducing the size of MOFs to the nanometer scale (denoted as NMOFs) has attracted considerable attention in recent years as NMOFs possess superior properties compared to their macroscopic counterparts, which are useful for various applications including sensing [109]. NMOFs with various morphologies (e.g. nanospheres, nanocubes, nanotubes, nano-sheets and nanorods) have been pre-pared by various synthetic methods [110,111]. Among different nanostructured materials, nanotubes are particularly attractive and show unique properties as their tubular nanostructures provide access to three different contact regions, namely internal and external surfaces as well as both ends. Zhang *et al.* [112] fabricated highly luminescent MOFNTs, $[Cd_2(1,3,5\text{-}BTC)_2(H_2O)_2]$ (LMOF18), by adopting a rational self-sacrificing template strategy and combining the advantages of both ultrasonic synthesis and vapor diffusion technique. The fabricated MOFNTs with inner diameters of 50-150 nm and outer diameters of 100-300 nm have been successfully utilized as luminescent material for the detection of 2,4-DNT in vapor phase. Usually, 2,4-DNT is present in TNT samples as an impurity resulting from the manufacturing process. The equilibrium vapor pressure of 2,4-DNT (100 ppb) is 20 times higher than that of TNT (5 ppb). Thus, high vapor pressure makes 2,4-DNT the target molecule for TNT detection. The as-synthesized MOFNTs displayed strong solid state fluorescence centering at 406 nm in the range of 350-550 nm upon excitation at 315 nm at room temperature. To evaluate the detection ability of MOFNTs towards NACs, vapors of different nitro analytes and some representative non-nitro compounds, such as NB, 2-NT, 4-NT and 2,4-DNT as well as toluene, chloroform, DMF, benzene, acetonitrile, DMSO and THF respectively were exposed to the glass coated luminescent material. Strong fluorescence quenching effects were observed, which clearly revealed that the fluorescence properties of the MOFNTs were highly sensitive to all nitro analytes. The vapors of non-nitro organic solvents showed no obvious fluorescence quenching. Among various nitro analytes, 2,4-DNT vapors quenched the fluorescence intensity by 72.5% within 10 s. The LOD of the as-synthesized MOFNTs for fluorescence sensing of 2,4-DNT was calculated as low as 18.1 ppb. The MOFNTs showed good recyclability up to ten cycles, and after fluorescence spectroscopic titration with 2,4-DNT, the fluorescence performance recovered by heating the sample at 80 °C for about 2 min.

Apart from the vapor phase detection of nitro compounds, numerous luminescent MOFs have also been applied for the investigation of NACs in solution phase. Recently, our group synthesized a cadmium based LMOF $\{[Cd(IPA)(3\text{-}PCNH)]\}_n$ (LMOF19) by employing mechano-chemical and conventional synthetic methods [113]. The 2D luminescent MOF was constructed by using isophthalate and neutral N-donor ligand (3-PCNH). LMOF19 exhibited good chemical stability in various organic solvents which was established by soaking the samples in the respective media for 7 days and subsequent confirmation by the PXRD analysis. Taking the advantage of water stability, the sensing experiments for nitro aromatic molecules were carried out in aqueous media. The aqueous solution of LMOF exhibited strong emission profile near 430 nm on excitation at 295 nm at room temperature. Furthermore, fluorescence-quenching titrations were performed with incremental addition of different nitro analytes into aqueous suspensions of LMOF. Among the large pool of nitro analytes (TNP, 2,6-DNT, 2,4-DNT, 3-NT, 2-NT, 4-NT, NB, NM and DMNB), TNP has been selectively detected with 79% quenching percentage at 200 µL of TNP (2 mM) solution. Fluorescence quenching efficiency for TNP was calculated by the S-V equation (1.52×10^4 M^{-1}), and the material could detect as low as 14 ppb (0.06 µM) TNP in aqueous medium. Another highly luminescent cadmium based MOF, [Cd(5-BrIP)(TIB)]$_n$ (LMOF20), was reported by our group for the detection of TNP with increasing quenching ability [86]. The luminescent material showed high selectivity and efficiency towards TNP in water among the large range of nitro analytes. The quenching efficiency was observed to be 84% at 200 µL of TNP (2 mM) (Figure 9.10a-b). To explore the selective detection of TNP as well as the quenching mechanism, the analyses of S-V plots was carried out. The plot with TNP showed a non-linear pathway, whereas the other nitro analytes followed linear trends in S-V plots. The linear fitting of S-V plot revealed the quenching constant for TNP as 2.68×10^4 M^{-1}. The non-linear trend in the S–V plot is suggestive of the energy transfer between TNP and the framework or the combination of the static and dynamic quenching mechanism [101]. These two quenching pathways can be differentiated by the lifetime of the sensor in the presence and absence of the quencher. The lifetime of the LMOF material remains unchanged due to the formation of a non-emissive ground state complex through the bonding between the fluorophore and quencher in the static mechanism. In the dynamic quenching mechanism, the lifetime reduces due to the electron transfer between the excited state fluorophore and quencher through diffusion-controlled collisions [101].

The lifetime values before and after the addition of TNP remained unchanged suggesting that the quenching process chiefly follows static mechanism (Figure 9.10c). Moreover, the selective quenching by TNP is also attributed to the electron and energy transfer mechanisms as well as the electrostatic interactions between the electron deficient analyte and electron rich framework (Figure 9.10d). The resonance energy transfer from the fluorophore of the organic ligand to the non-emitting analytes occurs when there is spectral overlap between the absorption spectrum of the analyte and the emission band of the luminescent material. Such an overlap enhances the quenching efficiency and sensitivity for a particular analyte depending on the extent of spectral overlap. In the current case, the absorption band of TNP shows the largest spectral overlap with the emission band of the luminesce material, while other nitro analytes have almost no overlap. The energy transfer is a long-range process, and, hence, the emission quenching by TNP is carried over by the surrounding aromatic moiety or fluorophores, amplifying the quenching response.

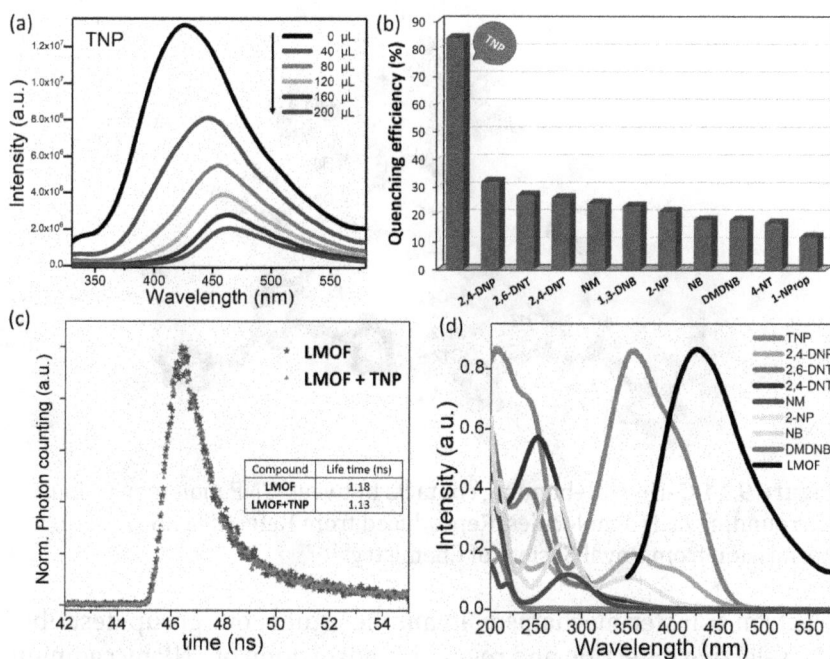

Figure 9.10 (a) Effect of [Cd(5-BrIP)(TIB)]$_n$ fluorescence spectra dispersed in water (2 mg/2 mL) upon incremental addition of aqueous solution of TNP (2 mM), (b) quenching efficiency of different nitro analytes, (c) spectral overlap between the absorption spectra of nitro analytes and the emission band of the LMOF and (d) fluorescence decay profile in the presence and absence of TNP. Reproduced from Reference 86 with permission from Royal Society of Chemistry.

The highly acidic phenolic oxygen as well as nitro group (–NO$_2$)

oxygens of TNP can participate in strong ionic/hydrogen bonding interactions with TIB as well as 5-BrIP linkers of the luminescent framework. Particularly, in this context, C–H···O hydrogen bonding interactions in the binary co-crystal between TIB and TNP supports the quenching process to a large extent (Figure 9.11). Overall, the non-linearity in the S–V plot and the spectral overlap of TNP advocate the energy transfer and the electron transfer mechanisms which are responsible for the selective fluorescence quenching with TNP; fluorescence quenching by other nitro analytes happens only through the

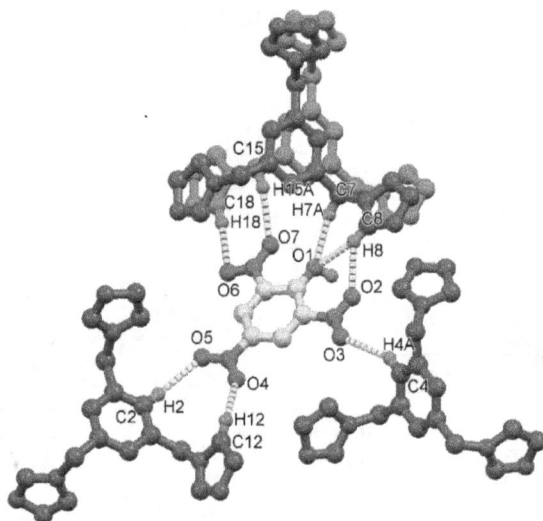

Figure 9.11 C–H···O H–bonding contacts between TNP moiety with four surrounding of TIB molecules. Reproduced from Reference 86 with permission from Royal Society of Chemistry.

electron transfer mechanism. In another study, our group described the selective detection and reversible adsorption of TNP by cadmium based luminescent MOF in aqueous medium [114]. The 3D complex $[Cd_3(SDB)_3(TIB)](H_2O)_2(1,4\text{-dioxane})(G)_x\}_n$ (LMOF21) was constructed by TIB and SDB ligands. The complex exhibited strong luminescence centered at 442 nm upon excitation at 320 nm (Figure 9.12a). The time resolved emission measurements showed good fluorescence behavior with mean lifetime (τ_m) of 3.812 ns ($\lambda_{ex} = 340$ nm). Luminescence study in different solvents revealed sensitivity for NB with ($-NO_2$) functionality. Efficient, fast and selective turn off in emission intensity (up to 95%) was observed for TNP in aqueous solution. The calculated fluorescence quenching efficiency for TNP by fitting exponential quenching equation was 2.43×10^4 M^{-1} ($R^2 = 0.99$) (Figure 9.12b). LMOF21 could detect the traces of TNP as low as 35 ppb (0.15 μM). The column chromatographic filler based on LMOF displayed good adsorption capacity for TNP from aqueous solution, which was confirmed by the UV-Vis spectrophotometer (Figure 9.12c-d).

In general, cadmium based luminescent materials are limited for practical use when compared to the zinc based LMOFs. However, cadmium based LMOFs with unique topologies and sensing properties are useful for fundamental studies, thus, opening the pathways for other approaches.

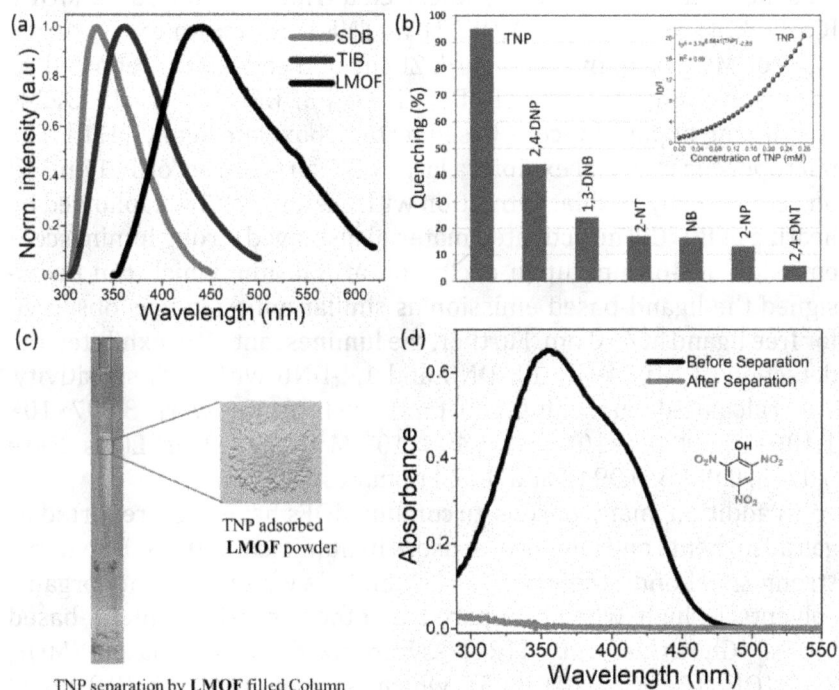

Figure 9.12 (a) Photoluminescence spectra of LMOF21 (λ_{ex} = 320 nm, λ_{em} = 442 nm), SDB and TIB ligands in solid state, (b) percentages of fluorescence quenching upon addition of different nitro analytes (280 μL) (Inset: non-linear S-V plot for TNP), (c) digital photographs depicting TNP separation from aqueous solution by the chromatographic column and (d) UV–Vis spectra of aqueous solutions of TNP (10^{-4} M) before and after separation. Reproduced from Reference 114 with permission from American Chemical Society.

9.3.3 Other Metal-based LMOFs as Sensing Materials for Nitro Explosive Detection

Apart from zinc and cadmium LMOFs, several other transition, lanthanide and main group metal based MOFs also display high selectivity for the detection of nitro explosives [115-118]. Cui *et al.* [115] reported a Cu(I) based coordination polymer, $[Cu_4Br(CN)(mtz)_2]_n$ (LMOF22), obtained from the *in-situ* formation of Cu(I) from Cu(II) and ring to ring conversion of ligand. The 3D luminescent material exhibits a highly sensitive colorimetric performance for NB and 2-NT. Highly conjugated double bond containing ligands show excellent fluorescence properties even with paramagnetic transition metal ions. Recently, Berke *et al.* synthesized a triptycene based Co-MOFs, $[Co_2(TTC)(DMF)(H_2O)]$ (LMOF23), as TNP sensor with the K_{SV} value 5.2×10^4 M^{-1} [119]. In this regard, Zhang and coworkers reported an interesting Cu(II) ($[Cu_3(TBPD)_{1.5}(H_2O)_3] \cdot 3DEF \cdot 20H_2O$) (LMOF24) comprising of pentiptycene based tertacarboxylate ligand [120]. The microporous material exhibits a large BET surface area of 1117 m^2g^{-1} on activation by Soxhlet extraction with acetone for 36 h, followed by heating at 80 °C. The activated material displayed strong luminescent emission at 465 nm, upon excitation at 330 nm, which can be assigned the ligand-based emission as similar emission was observed for free ligand at 418 nm. Further, the luminescent MOF exhibited the detection of NB, 4-NP, 1,3-DNB and 1,4-DNB with high sensitivity. The calculated quenching constants such as K_{SV} were 3.097×10^6, 1.406×10^6, 4.420×10^5 and 1.498×10^5 M^{-1} as well as LODs were 0.0775, 0.0896, 0.0949 and 0.123 ppm, respectively.

In addition, many porous zirconium MOFs have been reported as gas adsorbents and chemical sensors in humid conditions. Due to the strong Zr-O bonds, their chemical stability in water and other organic solvents is high when compared to other transition metal-based MOFs. Ghosh *et al.* [121] synthesized a Zr based MOF, $Zr_6O_4(OH)_4(PPDC)_6$ (LMOF25), which is composed of Zr(IV) metal cations and remains highly stable in water. The sizes of the pore windows is 11.5 and 23 Å, which are larger than the size of the nitro analytes. It could permit easy diffusion of analytes through the open pores, keeping the electron rich MOF and electron deficient nitro analytes in close contact. Studies disclosed that nitro analytes, such as 2,4-DNT, TNP and TNT, could change the fluorescent intensity of the guest free MOF dispersed in water. Fluorescence quenching could be clearly perceived for TNP concentrations as low as 2.6 µM, while other nitro analytes had minor effects on the intensity of luminescent MOF. The linear fitting of the S-V plot for TNP gave the quenching constant of 2.9×10^4 M^{-1}. This complex demonstrates highly selective and sensitive detection of TNP in water even in the presence of competing nitro analytes. Ghosh *et al.* [122] further synthesized a 3D MOF matrix: $Zr_6O_4(OH)_4(atdc)_6$ (LMOF26) with a pendant amine functionality for selective sensing of TNP in the aqueous phase, with increasing detection limit for TNP as low as 0.4 ppm. Gu *et al.* [123] reported a long wavelength emission sensor for the detection of TNT in aqueous media. The porphyrin-based zirconium activated MOF showed BET surface area up to 1413 m^{2}g^{-1} and exhibited strong emission profile at 651 nm upon excitation at 590 nm. Aqueous solutions of LMOF detected TNT among the large group of nitro analytes (2,6-DNT, NB, 3,5-DNBA, 3-NBA, 4-NT, BA, CB, TNP). The quenching constant (K_{SV}) for TNT was 3.5×10^4 M^{-1}, and it could be sense TNT as low as 0.46 μM.

Due to the small stoke shifts in transition metals, detection by naked eye is difficult. Lanthanide MOFs, especially Tb^{3+} and Eu^{3+}, are unique as their luminescence, when excited at a particular wavelength, exhibits large stoke shifts [56]. Zhou and co-workers reported a highly efficient Eu-based luminescent MOF, $(Me_2NH_2)_3[Eu_3(MHFDA)_4(NO_3)_4(DMF)_2] \cdot 4H_2O \cdot 2MeCN$, for the detection of NACs [116]. The MOF was heated at 100-170 °C in a vacuum drying oven and subsequently soaked in DMF to obtain $Eu_3(H_{0.75}MHFDA)_4(NO_3)_4 \cdot 5.5DMF$ (LMOF27). It exhibited red luminescence when excited at 340 nm, with characteristic emissions at 579, 591 and 614 nm with shoulder peaks at 619, 652 and 700 nm in the emission spectrum, corresponding to $5D0 \rightarrow 7FJ$ (J=0–4) transitions of Eu^{3+}, respectively. The emission of the blue light in free MHFDA completely disappears, and the ligand as light harvesting fluorophores (antenna effect) effectively transfers energy to Eu^{3+} ions. LMOF27 is a potential luminescent material for NACs, especially for TNP. The emission was almost completely quenched when the concentration of TNP reached 1.0×10^{-3} M. In another study, Partha and coworkers reported a submicron/nano-sized metal-organic material: $[Y_2(PDA)_3(H_2O)]_2 \cdot H_2O$ (LMOF28) [117]. Doping of 10% terbium ions (Tb^{3+}) in place of Y^{3+} afforded the desired phosphor material: Tb@LMOF28. Upon activation, the obtained dehydrated material shows a highly intense visible green emission on exposure to UV light. The luminescence in acetonitrile shows high sensitivity and a quick visual detection ability towards the presence of a trace of amount of TNP, DNB, DNT, etc., *via* luminescence quenching. The observed quenching constants (K_{sv}) were 70920, 44000, 35430 and 7690 M^{-1} for TNP, DNB, DNT and NB, respectively. The larger observed K_{sv} values revealed extremely high sensitivity, and Tb@LMOF28 was confirmed as one of the best sensitive luminescence-based metal-organic detector for TNP as compared to other NACs.

Recently, Moon and coworkers reported main group metal (Li$^+$) based luminescent MOF, {Li$_3$[Li(DMF)$_2$](CPMA)$_2$}·4DMF·H$_2$O (LMOF29), for the detection of nitro aromatic based explosives [124]. SBU contains 1D chains with tetrahedrally coordinated LiO$_4$. Each chain is further bridged by four neighboring chains of dicarboxylate ligands in four different directions which leads to a 3D framework with two kinds of rectangular pores, along the c-axis. The 1D channels filled with free and coordinated DMF molecules were evacuated by thermal treatment. The activated material was investigated as a sensor for nitro analytes upon soaking 2,4-DNT, NB, TO and BZ liquids (a concentrated DMF solution used for 2,4-DNT). A significant color change was observed from yellow to red or deep orange for NB and 2,4-DNT analytes, respectively (Figure 9.13a). In contrast, the color remains unchanged upon soaking in TO and BZ. The origin of such

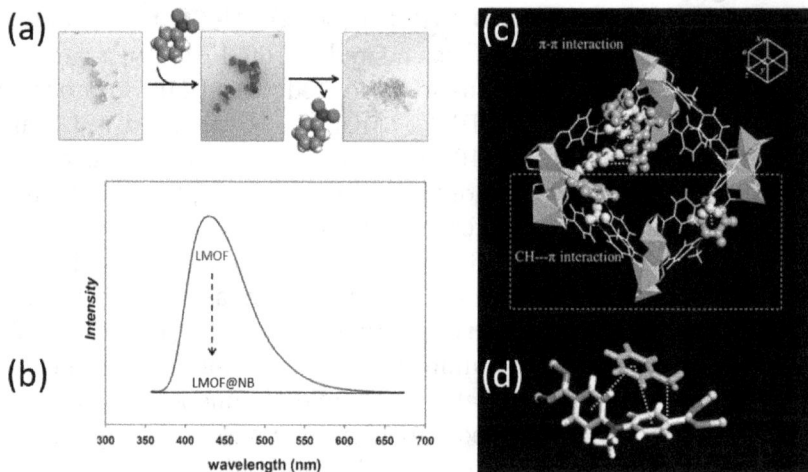

Figure 9.13 Detection of nitrobenzene in LMOF29. (a) Digital photographs illustrating the colorimetric detection of NB, and regenerated by heating, (b) effect on the emission spectra of activated LMOF upon exposure to the NB vapors and (c-d) the representative π-π and C-H···π interactions are observed in X-ray structure of LMOF29@nitrobenzene and close-up shot of the interaction between nitrobenzene and CPMA ligand. Reproduced from reference 124 with permission from American Chemical Society.

color change in NB was investigated using SXRD, along with IR and elemental analysis. The SXRD data revealed that strong π-π interactions of the aromatic pore surface facilitated the adsorption of NB inside the channels as a guest molecules. The significant color changes in the structure were amplified by the strong π-π and C–H···π interactions between the guest -NO₂ group and the linker π system (Figure 9.13c-d). The UV-vis absorption spectrum displayed two absorption peaks, at 248 and 352 nm, with appearance of an additional absorption peak at around 500 nm for as-synthesized material in the solid state. The interactions between the electron-rich linker and the electron deficient NB could be attributed to the charge-transfer between the aromatic rings. Such charge transfer was the reason behind the intense red color material. The material could be regenerated by heating under vacuum at 100 °C for 5 minutes. The as-synthesized LMOF exhibits intense fluorescence at λ_{max} = 430 nm upon excitation at 345 nm, and the donor N-methylamino group as well as acceptor carboxylate groups within the linker result in strong intra-ligand charge transfer (ILCT). On the other hand, NB and DNT soaked materials are completely non-emissive under excitation at 345 nm (Figure 9.13b) which can be explained due to the electron transfer donor-acceptor mechanism *via* host-guest interaction.

9.3.4 Detection by MOF based Luminescent Paper Strips

For effective and rapid detection of NACs, LMOF based luminescent paper strips are very convenient in the area of forensic and analytical sciences. The fabrication of luminescent paper strips can also be easily achieved by dip-coating method. Suresh *et al.* [103] fabricated luminescent paper strips by exploiting [Zn(μ₂-1Hade)(μ₂-SO₄)] (LMOF13) luminescent coordination polymer. The strips were generated by dipping Whatman filter paper strips in MOF dispersed ethanolic solution followed by drying in desiccator for one day. The test strips emitted a light-blue color under 365 nm UV light which were further partially dipped in aqueous solutions of nitro aromatic compounds. The paper strips after treatment with analyte solutions under 365 nm UV light showed a selective sensing for 2,4-DNP and TNP nitro analytes, as shown in Figure 9.14. Li *et al.* [125] reported rhodamine 6G immobilized europium based luminescent MOF RGH–Eu(1,3,5-BTC) (LMOF30), which was further utilized for the fabrication of luminescent paper strips. Test strips were prepared by dipping Whatman filter paper strips in RGH–Eu(1,3,5-BTC) concentrated suspensions (5 mg/mL), followed by drying in air. Various concentrations of TNP were added onto these test strips dropwise, and a gradual colorimetric change from colorless to orange was observed in daylight. These potential results demonstrate that RGH–Eu(1,3,5-BTC) is sensitive towards TNP, which may provide a rapid and portable paper strip detection method to recognize TNP with naked eye.

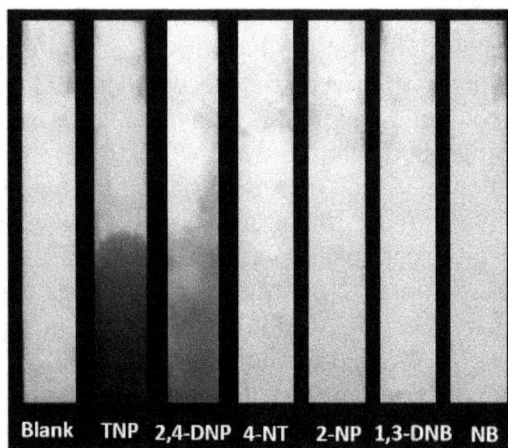

Figure 9.14 Digital images of LMOF13 test strips after they were dipped in aqueous solutions of nitroaromatic compounds (2 mM). Reproduced from References 103 with permission from American Chemical Society.

9.4 Conclusions

Recent developments in the area of luminescent metal organic frameworks and their application as a potential sensor material for the detection of organic nitro aliphatic and aromatic compounds are the highlighted in this chapter. Judicious choice of the metal nodes and the conjugated/functionally decorated ligand moieties can provide the required attributes such as porosity, luminescence, etc., in MOFs for sensing/detection of analytes. Luminescence property of the MOFs is exploited for the sensing and detection of the electron deficient NOCs and NACs by fluorescent quenching. General synthetic strategies for the design and synthesis of LMOFs and theoretical aspects applicable to the origin of luminescence in LMOFs are also discussed. Mainly focused in this chapter are Zn(II) and Cd(II) based LMOFs for sensing and detection of hazardous explosives (NOCs and

NACs). A brief account of the other metal-based LMOFs, with Ln based MOF in particular, as sensing materials and LMOF coated paper strips for naked eye detection of the explosives with distinct color change is also presented. The development of efficient LMOFs with high stability as well as selective and sensitive sensing with rapid detection response towards NOCs/NACs in aqueous phase for applications in homeland security and environment still remains a challenge and necessitates further research.

List of Abbreviations

EPA	Environmental Protection Agency
VOS	volatile organic solvents
ONCs	organic nitro compounds
1D	one dimensional
2D	two dimensional
3D	three dimensional
PXRD	powder X-ray diffraction
SXRD	single crystal X-ray diffraction
IR	infrared
PL	photoluminescence
ppm	parts per million
ppb	parts per billion
MOFNTs	MOF nanotubes
LOD	limit of detection
BET	Brunauer-Emmett-Teller
CB	conduction band
LUMO	lowest unoccupied molecular orbital
HOMO	highest occupied molecular orbital
bpdc	4,4'-biphenyldicarboxylate
bpee	1,2-bipyridylethene
BTOT	4,4',4''-(benzene-1,3,5-triyltris (oxy)tribenzoic acid
1,2,4-BTC	1,2,4-benzenetricarboxylate
pdbs	1,2-phenylendiamine-N,N'-bis(3-tert-butyl-5-(carboxyl)-salicylide-ene
4,4'-bpy	4,4'-bipyridine
SDB	4,4'-sulfonyl dicarboxylate
oba	4,4'-oxybis-(benzoic acid)
PPZ	piperazine
H$_4$tcpb	1,2,4,5-tetrakis(4-carboxyphenyl)benzene
bpeb	1,4-bis[2-(4-pyridyl)ethenyl]benzene

1,3,5-BTC	1,3,5-benzenetricarboxylate
IPA	isophthalic acid
TPA	terephthalic acid
1,4-BDC	1,4-benzene dicarboxylate
1Hade	1H-adenine
Ade	adenine
TIPA	tri(4-imidazolylphenyl)amine
HCOP	hexa[4-(carboxyphenyl)oxamethyl]-3-oxapentane acid
5-BrIP	5- bromo isophthalic acid
H_4TTC	1,4,5,8-triptycenetetracarboxylic acid
4-PCIH	4-pyridylcarboxaldehyde isonicotinoylhydrazone
3-PCNH	3-pyridylcarboxaldehyde nicotinoylhydrazone
mtz	5-methyl tetrazole
TBPD	5,5'-((5,7,12,14-tetrahydro-5,14:7,12-bis([1,2]ben-zeno)pentacene-6,13-diyl)bis(ethyne-2,1-diyl))diisophthalic acid
PPDC	2-phenylpyridine-5,4-dicarboxylic acid
atdc	2'-amino-[1,1':4',1"-terphenyl]-4,4"-dicarboxylate
H_2MHFDA	9-methyl-9-hydroxy-fluorene-2,7-dicarboxylic acid
PDA	1,4-phenylenediacetate
H_2CPMA	bis(4-carboxyphenyl)-N-methylamine
2,2'-bpy	2,2'-bipyridine
2,2'-dmbpy	2,2'-dimethyl-4,4'bipyridine
TIB	1,3,5-tris(imidazol-1-ylmethyl)benzene
RGH	rhodamine 6G hydrazide
DEF	N,N'-diethylformamide
DMA	N,N'-dimethylacetamide
DMSO	dimethyl sulfoxide
THF	tetrahydrofuran
DCM	dicloromethane
MeCN	acetonitrile
BZ	benzene
TO	toluene
NB	nitrobenzene
NM	nitro methane
2-NP	2-nitrophenol
3-NP	3-nitrophenol
4-NP	4-nitrophenol
4-NT	4-nitrotoluene
3-NT	3-nitrotoluene

2-NT	2-nitrotoluene
1,3-DNB	1,3-dinitrobenzene
1,4-DNB	1,4-dinitrobenzene
2,4-DNP	2,4-dinitrophenol
TNP	2,4,6-trinitrophenol
2,4-DNT	2,4-dinitrotoluene
2,6-DNT	2,6-dinitrotoluene
TNT	2,4,6-trinitrotoluene
DMNB	2,3-dimethyl-2,3-dinitrobutane
TNG	nitroglycerin
TNP	2,4,6-trinitro phenol
HMX	cyclotetramethylenetetranitramine
RDX	cyclotrimethylenetrinitramine
3,5-DNBA	3,5-dinitrobenzoic acid
3-NBA	3-nitrobenzoic acid
PA	picric acid
BA	benzoic acid
CB	chlorobenzene

References

1. Toal, S. J., and Trogler, W. C. (2006) Polymer sensors for nitroaromatic explosives detection. *Journal of Materials Chemistry*, **16**, 2871-2883.
2. Senesac, L., and Thundat, T. G. (2008) Nanosensors for trace explosive detection. *Materials Today*, **11**, 28-36.
3. Salinas, Y., Martinez-Manez, R., Marcos, M. D., Sancenon, F., Costero, A. M., Parra, M., and Gil, S. (2012) Optical chemosensors and reagents to detect explosives. *Chemical Society Reviews*, **41**, 1261-1296.
4. McQuade, D. T., Pullen, A. E., and Swager, T. M. (2000) Conjugated polymer-based chemical sensors. *Chemical Reviews*, **100**(7), 2537-2574.
5. Lan, A. J., Li, K. H., Wu, H. H., Olson, D. H., Emge, T. J., Ki, W., Hong, M. C., and Li, J. (2009) A luminescent microporous metal-organic framework for the fast and reversible detection of high explosives. *Angewandte Chemie, International Edition*, **48**(13), 2334-2338.
6. Pramanik, S., Zheng, C., Zhang, X., Emge, T. J., and Li, J. (2011) New microporous metal–organic framework demonstrating unique selectivity for detection of high explosives and aromatic compounds. *Journal of the American Chemical Society*, **133**(12), 4153-4155.
7. Germain, M. E., and Knapp, M. J. (2009) Optical explosives detection: from color changes to fluorescence turn-on. *Chemical Society Revi-*

ews, **38**(9), 2543-2555.

8. Moore, D. S. (2004) Instrumentation for trace detection of high explosives. *Review of Scientific Instruments,* **75**, 2499-2512.

9. Thomas, S. W., Joly, G. D., and Swager, T. M. (2007) Chemical sensors based on amplifying fluorescent conjugated polymers. *Chemical Reviews,* **107**(4), 1339-1386.

10. Liu, B. (2012) Metal-organic framework-based devices: separation and sensors. *Journal of Materials Chemistry,* **22**, 10094-10101.

11. Hakansson, K., Coorey, R. V., Zubarev, R. A., Talrose, V. L., and Hakansson, P. (2000) Low mass ions observed in plasma desorption mass spectrometry of high explosives. *Journal of Mass Spectrometry,* **35**, 337-346.

12. Walsh, M. E. (2001) Determination of nitroaromatic, nitramine, and nitrate ester explosives in soil by gas chromatography and an electron capture detector. *Talanta,* **54**(3), 427-438.

13. Sylvia, J. M., Janni, J. A., Klein, J. D., and Spencer, K. M. (2000) Surface-enhanced raman detection of 2,4-dinitrotoluene impurity vapor as a marker to locate landmines. *Analytical Chemistry,* **72**(23), 5834-5840.

14. Yinon, J. (1982) Mass spectrometry of explosives: Nitro compounds, nitrate esters, and nitramines. *Mass Spectrometry Reviews,* **1**, 257-307.

15. Mathurin, J. C., Faye, T., Brunot, A., and Tabet, J. C. (2000) High-pressure ion source combined with an in-axis ion trap mass spectrometer. 1. Instrumentation and applications. *Analytical Chemistry,* **72**(20), 5055-5062.

16. Hallowell, S. F. (2001) Screening people for illicit substances: A survey of current portal technology. *Talanta,* **54**(3), 447-458.

17. Vourvopoulos, G., and Womble, P. C. (2001) Pulsed fast/thermal neutron analysis: A technique for explosives detection. *Talanta,* **54**(3), 459-468.

18. Krausa, M., and Schorb, K. (1999) Trace detection of 2,4,6-trinitrotoluene in the gaseous phase by cyclic voltammetry. *Journal of Electroanalytical Chemistry,* **461**, 10-13.

19. Wallis, E., Griffin, T. M., Popkie, Jr., N., Eagan, M. A., McAtee, R. F., Vrazel, D., and McKinly, J. (2005) Instrument response measurements of ion mobility spectrometers in situ: Maintaining optimal system performances of fielded systems. *Proceedings of the SPIE-The International Society for Optical Engineering,* **5795**, 54-64.

20. Eiceman, G. A., and Stone, J. A. (2004) New uses of previously unheralded analytical instruments. *Analytical Chemistry,* **76**(21), 390A-397A.

21. Furton, K. G., and Myers, L. J. (2001) The scientific foundation and efficacy of the use of canines as chemical detectors for explosives. *Talanta,* **54**(3), 487-500.

22. Eliseeva, S. V., and Bunzli, J. C. G. (2010) Lanthanide luminescence for functional materials and bio-sciences. *Chemical Society Reviews* **39**, 189-227.

23. Binnemans, K. (2009) Lanthanide-Based Luminescent Hybrid Materials. *Chemical Reviews,* **109**(9), 4283-4374.

24. Carlos, L. D., Ferreira, R. A. S., de Zea Bermudez, V., Julian-Lopez, B., and Escribano, P. (2011) Progress on lanthanide-based organic-inorganic hybrid phosphors. *Chemical Society Reviews,* **40**, 536-549.

25. Hwang, S. H., Moorefield, C. N., Newkome, G. R. (2008) Dendritic macromolecules for organic light-emitting diodes. *Chemical Society Reviews,* **37**, 2543-2557.

26. Lo, S. C., and Burn, P. L. (2007) Development of dendrimers: Macromolecules for use in organic light-emitting diodes and solar cells. *Chemical Reviews,* **107**(4), 1097-1116.

27. Grimsdale, A. C., Leok Chan, K., Martin, R. E., Jokisz, P. G., and Holmes, A. B. (2009) Synthesis of light-emitting conjugated polymers for applications in electroluminescent devices. *Chemical Reviews,* **109**(3), 897-1091.

28. Veinot, J. G. C., and Marks, T. J. (2005) Toward the ideal organic light-emitting diode. The versatility and utility of interfacial tailoring by cross-linked siloxane interlayers. *Accounts of Chemical Research,* **38**(8), 632-643.

29. Thomas, S. W., Amara, J. P., Bjork, R. E., and Swager, T. M. (2005) Amplifying fluorescent polymer sensors for the explosives taggant 2,3-dimethyl-2,3-dinitrobutane (DMNB). *Chemical Communications,* 4572-4574.

30. Nie, H. R., Zhao, Y., Zhang, M., Ma, Y. G., Baumgarten M., and Mullen, K. (2011) Detection of TNT explosives with a new fluorescent conjugated polycarbazole polymer. *Chemical Communications,* **47**, 1234-1236.

31. Liu, Y., Mills, R. C., Boncella, J. M., and Schanze, K. S. (2001) Fluorescent polyacetylene thin film sensor for nitroaromatics. *Langmuir,* **17**(24), 7452-7455.

32. Sohn, H., Sailor, M. J., Magde, D., and Trogler, W. C. (2003) Detection of nitroaromatic explosives based on photoluminescent polymers containing metalloles. *Journal of the American Chemical Society,* **125**(13), 3821-3830.

33. Yang, J. S., and Swager, T. M. (1998) Fluorescent porous polymer films as TNT chemosensors: Electronic and structural effects. *Journal of the American Chemical Society,* **120**(46), 11864-11873.

34. Germain, M. E., Vargo, T. R., Khalifah, P. G., and Knapp, M. J. (2007) Fluorescent detection of nitroaromatics and 2,3-dimethyl- 2,3-dinitrobutane (DMNB) by a zinc complex: (salophen)Zn. *Inorganic Chemistry,* **46**(11), 4422-4429.

35. Zhou, H. C., Long, J. R., and Yaghi, O. M. (2012) Introduction to metal-

organic frameworks. *Chemical Reviews,* **112**(2), 673-674.

36. Batten, S. R., Champness, N. R., Chen, X. M., Garcia-Martinez, J., Kitagawa, S., Öhrström, L., O'Keeffe, M., Suh, M. P., and Reedijk, J. (2012) Coordination polymers, metal-organic frameworks and the need for terminology guidelines. *CrystEngComm,* **14**, 3001-3004.

37. Tranchemontagne, D.   J.,   Mendoza-Cortés, J.   L.,   O'Keeffe, M., and Yaghi, O. M. (2009) Secondary building units, nets and bonding in the chemistry of metal-organic frameworks, *Chemical Society Reviews,* **38**, 1257-1283.

38. Farha, O. K., Eryazici, I., Jeong, N. C., Hauser, B. G., Wilmer, C. E., Sarjeant, A.A., Snurr, R.Q., Nguyen, S. T., Yazaydın, A. Ö., and Hupp, J. T. (2012) Metal–organic framework materials with ultrahigh surface areas: Is the sky the limit? *Journal of the American Chemical Society,* **134**(36), 15016-15021.

39. Wu, H., Gong, Q., Olson, D. H., and Li, J. (2012) Commensurate adsorption of hydrocarbons and alcohols in microporous metal organic frameworks. *Chemical Reviews,* **112**(2), 836-868.

40. Poloni, R., Lee, K., Berger, R. F., Smit B., and Neaton, J. B. (2014) Understanding trends in CO_2 adsorption in metal-organic frameworks with open-metal sites. *The Journal of Physical Chemistry Letters,* **5**(5), 861-865.

41. Corma, A., Garcia, H., and Llabres I Xamena, F. X. (2010) Engineering metal organic frameworks for heterogeneous catalysis. *Chemical Reviews,* **110**(8), 4606-4655.

42. Ramaswamy, P., Wong, N. E., and Shimizu, G. K. H. (2014) MOFs as proton conductors - challenges and opportunities. *Chemical Society Reviews,* **43**, 5913-5932.

43. Shimizu, G. K. H., Taylor, J. M., Kim, S. (2013) Proton conduction with metal-organic frameworks. *Science,* **341**(6144), 354-355.

44. Liana, X., and Yan, B. (2016) A lanthanide metal-organic framework (MOF-76) for adsorbing dyes and fluorescence detecting aromatic pollutants. *RSC Advances,* **6**, 11570-11576.

45. Fang, X., Zong, B. and Mao, S., (2018) Metal-organic framework-based sensors for environmental contaminant sensing. *Nano-Micro Letters,* **10**(4), 64.

46. Hu, Z., Deibert, B. J., and Li, J. (2014) Luminescent metal-organic frameworks for chemical sensing and explosive detection. *Chemical Society Reviews,* **43**, 5815-5840.

47. Banerjee, D., Hu, Z., and Li, J. (2014) Luminescent metal-organic frameworks as explosive sensors. *Dalton Transactions,* **43**, 10668-10685.

48. Bobbitt, N. S., Mendonca, M. L., Howarth, A. J., Islamoglu, T., Hupp, J. T., Farha, O. K., Snurr, R. Q. (2017) Metal–organic frameworks for the removal of toxic industrial chemicals and chemical warfare agents. *Chemical Society Reviews,* **46**, 3357-3385.

49. Du, T., Zhao, C., Rehman, F., Lai, L., Li, X., Sun, Y., Luo, S., Jiang, H., Gu, N., Selke, M., and Wang, X. (2017) In situ multimodality imaging of cancerous cells based on a selective performance of Fe^{2+}-adsorbed zeolitic imidazolate framework-8. *Advanced Functional Materials*, **27**(5), 1603926.

50. Cai, W., Gao, H., Chu, C., Wang, X., Wang, J., Zhang, P., Lin, G., Li, W., Liu, G., and Chen, X. (2017) Engineering phototheranostic nanoscale metal-organic frameworks for multimodal imaging-guided cancer therapy. *ACS Applied Materials & Interfaces*, **9**(3), 2040-2051.

51. Horcajada, P., Gref, R., Baati, T., Allan, P. K., Maurin, G., Couvreur, P., Férey, G., Morris, R. E., Serre, C. (2012) Metal-organic frameworks in biomedicine. *Chemical Reviews*, **112**, 1232-1268.

52. Chen, W., Zhuang, Y., Wang, L., Lv, Y., Liu, J., Zhou, T. L., and Xie, R. J. (2018) Color-tunable and high-efficiency dye-encapsulated metal-organic framework composites used for smart white-light-emitting diodes. *ACS Applied Materials & Interfaces*, **10**(22), 18910-18917.

53. Yu, J., Cui, Y., Xu, H., Yang, Y., Wang, Z., Chen, B., and Qian, G. (2013) Confinement of pyridinium hemicyanine dye within an anionic metal-organic framework for two-photon-pumped lasing. *Nature Communications*, **4**, 2719.

54. Ju, K. S., and Parales, R. E., (2010) Nitroaromatic compounds, from Synthesis to biodegradation. *Microbiology and Molecular Biology Reviews*, **74**(2), 250-72.

55. Allendorf, M. D., Bauer, C. A., Bhaktaa, R. K., and Houka, R. J. T. (2009) Luminescent metal-organic frameworks. *Chemical Society Reviews*, **38**, 1330-1352.

56. Cui, Y., Yue, Y., Qian, G., and Chen, B. (2012) Luminescent functional metal-organic frameworks. *Chemical Reviews*, **112**, 1126-1162.

57. Zhu, S. Y., and Yan, B. (2018) A novel covalent post-synthetically modified MOF hybrid as a sensitive and selective fluorescent probe for Al^{3+} detection in aqueous media. *Dalton Transactions*, **47**, 1674-1681.

58. Sharma, S., and Ghosh, S. K. (2018) Metal-organic framework-based selective sensing of biothiols via chemidosimetric approach in water. *ACS Omega*, **3**(1), 254-258.

59. Shustova, N. B., McCarthy, B. D., and Dinca, M. (2011) Turn-on fluorescence in tetraphenylethylene-based metal-organic frameworks: An alternative to aggregation-induced emission. *Journal of the American Chemical Society*, **133**(50), 20126-20129.

60. Zhang, M., Zhang, L., Xiao, Z., Zhang, Q., Wang, R., Dai, F., and Sun, D. (2016) Pentiptycene-based luminescent Cu (II) MOF exhibiting selective gas adsorption and unprecedentedly high-sensitivity detection of nitroaromatic compounds (NACs). *Scientific Reports*, **6**, 20672.

61. Dai, J.-C., Wu, X.-T., Fu, Z.-Y., Hu, S.-M., Du, W.-X., Cui, C.-P., Wu, L.-M.,

Zhang, H.-H., and Sun, R.-Q. (2002) Synthesis, structure, and fluorescence of the novel cadmium(II)-trimesate coordination polymers with different coordination architectures. *Chemical Communications*, **41**(6),1391-1396.

62. Chen, W., Wang, J.-Y., Chen, C., Yue, Q., Yuan, H.-M., Chen, J.-S., and Wang, S.-N. (2003) Photoluminescent metal-organic polymer constructed from trimetallic clusters and mixed carboxylates. *Inorganic Chemistry*, **42**(4), 944-946.

63. Khatua, S., Goswami, S., Biswas, S., Tomar, K., Jena, H. S., and Konar, S. (2015) Stable multiresponsive luminescent MOF for colorimetric detection of small molecules in selective and reversible manner. *Chemistry of Materials*, **27**, 5349–5360.

64. Bisht, K. K., Kathalikkattil, A. C., Suresh, E., (2012) Structure modulation, argentophilic interactions and photoluminescence properties of silver (I) coordination polymers with isomeric N-donor ligands. *RSC Advances*, **2**, 8421-8428.

65. Einkauf, J. D., Kelley, T. T., Chan, B. C., and de Lill, D. T. (2016) Rethinking sensitized luminescence in lanthanide coordination polymers and MOFs: Band sensitization and water enhanced Eu luminescence in $[Ln(C15H9O5)_3(H_2O)_3]_n$ (Ln = Eu, Tb). *Inorganic Chemistry*, **55**(16), 7920-7927.

66. Dolgopolova, E. A., Ejegbavwo, O. A., Martin, C. R., Smith, M. D., Setyawan, W., Karakalos, S. G., Henager, C. H., zur Loye, H. C., and Shustova, N. B. (2017) Multifaceted modularity: A key for stepwise building of hierarchical complexity in actinide metal–organic frameworks. *Journal of the American Chemical Society*, **139**, 16852–1686.

67. Chandrasekhar, P., Mukhopadhyay, A., Savitha, G., and Moorthy, J. N. (2016) Remarkably selective and enantiodifferentiating sensing of histidine by a fluorescent homochiral Zn-MOF based on pyrene-tetralactic acid. *Chemical Science*, **7**, 3085-3091.

68. Lu G., and Hupp, J. T. (2010) Metal–organic frameworks as sensors: A ZIF-8 based Fabry–Pérot device as a selective sensor for chemical vapors and gases. *Journal of the American Chemical Society*, **132**(23), 7832-7833.

69. Manna, B., Desai A. V., and Ghosh, S. K (2016) Neutral N-donor ligand based flexible metal-organic frameworks. *Dalton Transactions*, **45**, 4060-4072.

70. Kreno, L. E., Leong, K., Farha, O. K., Allendorf, M., Duyne, R. P. V., and Hupp, J. T. (2012) Metal-organic framework materials as chemical sensors. *Chemical Reviews*, **112** (2), 1105-1125.

71. Lin, C. K., Zhao, D., Gao, W.Y., Yang, Z., Ye, J., Xu, T., Ge, Q., Ma, S., and Liu, D. J. (2012) Tunability of band gaps in metal-organic frameworks. *Inorganic Chemistry*, **51**(16), 9039-9044.

72. Sun, R., Li, Y. Z., Bai, J., and Pan, Y. (2007) Synthesis, structure, water-induced reversible crystal-to-amorphous transformation, and lumi-

nescence properties of novel cationic spacer-filled 3D transition metal supramolecular frameworks from N,N',N''-Tris(carboxymethyl)-1,3,5-benzenetricarboxamide. *Crystal Growth & Design*, **7**(5), 890-894.

73. Wang, X. L., Bi, Y. F., Lin, H. Y., and Liu, G. C. (2007) Three novel Cd(II) metal–organic frameworks constructed from mixed ligands of dipyrido[3,2-d:2',3'-f]quinoxaline and benzene-dicarboxylate: From a 1-D ribbon, 2-D layered network, to a 3-D architecture. *Crystal Growth & Design*, **7**(6), 1086-1091.

74. Tian, Z., Lin, J., Su, Y., Wen, L., Liu, Y., Zhu, H., and Meng, Q. J. (2007) Flexible ligand, structural, and topological diversity: Isomerism in $Zn(NO_3)_2$ coordination polymers. *Crystal Growth & Design*, **7**(9), 1863-1867.

75. Perry, IV, J. J., Perman, J. A., and Zaworotko, M. J. (2009) Design and synthesis of metal-organic frameworks using metal-organic polyhedra as supermolecular building blocks. *Chemical Society Reviews*, **38**, 1400-1417.

76. Gangu, K. K., Maddila, S., Mukkamala S. B., and Jonnalagadda, S. B. (2016) A review on contemporary metal-organic framework materials. *Inorganica Chimica Acta*, **446**, 61-74.

77. Seetharaj, R., Vandana, P.V., Arya, P., and Mathew, S. (2016) Dependence of solvents, pH, molar ratio and temperature in tuning metal organic framework architecture. *Arabian Journal of Chemistry*, doi:10.1016/j.arabjc.2016.01.003.

78. Friščić, T. (2012) Supramolecular concepts and new techniques in mechanochemistry: cocrystals, cages, rotaxanes, open metal-organic frameworks. *Chemical Society Reviews*, **41**, 3493-3510.

79. Parmar, B., Rachuri, Y., Bisht, K. K., Laiya, R., and Suresh, E. (2017) Mechanochemical and conventional synthesis of Zn(II)/Cd(II) luminescent coordination polymers: Dual sensing probe for selective detection of chromate anions and TNP in aqueous phase. *Inorganic Chemistry*, **56**(5), 2627-2638.

80. Rachuri, Y., Parmar, B., Bisht, K. K., and Suresh, E. (2017) Solvothermal self-assembly of Cd^{2+} coordination polymers with supramolecular networks involving N-donor ligands and aromatic dicarboxylates: synthesis, crystal structure and photoluminescence studies. *Dalton Transactions*, **46**, 3623-3630.

81. Babu, R., Roshan, R., Kathalikkattil, A. C., Kim, D. W., and Park, D. W. (2016) Rapid, microwave-assisted synthesis of cubic, three-dimensional, highly porous MOF-205 for room temperature CO_2 fixation via cyclic carbonate synthesis, *ACS Applied Materials & Interfaces*, **8**, 33723–33731.

82. Muller, U., Schubert, M., Teich, F., Putter, H., Schierle-Arndt, K., Pastre, J. (2006) Metal-organic frameworks - prospective industrial applications. *Journal of Materials Chemistry*, **16**, 626-636.

83. Son, W. J., Kim, J., Kim J., and Ahn, W. S. (2008) Sonochemical synthesis of MOF-5. *Chemical Communications*, 6336-6338.

84. Chen, B., Xiang, S., and Qian, G. (2010) Metal–organic frameworks with functional pores for recognition of small molecules. *Accounts of Chemical Research*, **43**(8), 1115-1124.

85. Li, J. R., Yu, J., Lu, W., Sun, L. B., Sculley, J., Balbuena P. B., and Zhou, H. C., (2013) Porous materials with pre-designed single-molecule traps for CO_2 selective adsorption. *Nature Communications*, **4**, 1538-1544.

86. Rachuri, Y., Parmar, B., Bisht, K. K., and Suresh, E. (2016) Mixed ligand two dimensional Cd(II)/Ni(II) metal organic frameworks containing dicarboxylate and tripodal N-donor ligands: Cd(II) MOF is an efficient luminescent sensor for detection of picric acid in aqueous media. *Dalton Transactions*, **45**, 7881-7892.

87. Joarder, B., Desai, A. V., Samanta, P., Mukherjee, S., and Ghosh, S. K. (2015) Selective and sensitive aqueous-phase detection of 2,4,6-Trinitrophenol (TNP) by an amine-functionalized metal-organic framework. *Chemistry: A European Journal* **21**, 965-969.

88. Chen, B., Yang, Y., Zapata, F., Lin, G., Qian, G., and Lobkovsky, E. B. (2007) Luminescent open metal sites within a metal-organic framework for sensing small molecules. *Advanced Materials*, **19**, 1693-1696.

89. Shustova, N. B., Cozzolino, A. F., Reineke, S., Baldo, M., and Dinca, M. (2013) Selective turn-on ammonia sensing enabled by high-temperature fluorescence in metal-organic frameworks with open metal sites. *Journal of the American Chemical Society*, **135**(36), 13326-13329.

90. Chen, B., Wang, L., Zapata, F., Qian G., and Lobkovsky, E. B. (2008) A luminescent microporous metal–organic framework for the recognition and sensing of anions. *Journal of the American Chemical Society*, **130**(21), 6718-6719.

91. Chaudhari, A. K., Nagarkar, S. S., Joarder, B., and Ghosh, S. K. (2013) A Continuous π-Stacked Starfish Array of Two-Dimensional Luminescent MOF for Detection of Nitro Explosives. *Crystal Growth & Design*, **13**(8), 3716-3721.

92. Hendon, C. H., Tiana, D., Fontecave, M., Sanchez, C., D'arras, L., Sassoye, C., Rozes, L., Mellot-Draznieks, C., and Walsh, A. (2013) Engineering the optical response of the titanium-MIL-125 metal-organic framework through ligand functionalization. *Journal of the American Chemical Society*, **135**(30), 10942-10945.

93. Nagarkar, S. S., Joarder, B., Chaudhari, A. K., Mukherjee, S., and Ghosh, S. K. (2013) Highly selective detection of nitro explosives by a luminescent metal-organic framework. *Angewandte Chemie, International Edition*, **52**, 2881-2885.

94. Han, Y., Sheng, S., Yang, F., Xie, Y., Zhao, Mi., and Li, J. R. (2015) Size-

exclusive and coordination-induced selective dye adsorption in a nanotubular metal-organic framework. *Journal of Materials Chemistry A*, **3**, 12804-12809.

95. Wei, W., Lu, R., Tang, S., and Liu, X. (2015) Highly cross-linked fluorescent poly(cyclotriphosphazene-co-curcumin)microspheres for the selective detection of picric acid in solution phase. *Journal of Materials Chemistry A*, **3**, 4604-4611.

96. Hu, Z., Pramanik, S., Tan, K., Zheng, C., Liu, W., Zhang, X., Chabal, Y. J., and Li, J. (2013) Selective, sensitive, and reversible detection of vapor-phase high explosives via two-dimensional mapping: A new strategy for MOF-based sensors. *Crystal Growth & Design*, **13**(10), 4204-4207.

97. Zhang, Z., Xiang, S., Rao, X., Zheng, Q., Fronczek, F. R., Qian, G., and Chen, B. (2010) A rod packing microporous metal-organic framework with open metal sites for selective guest sorption and sensing of nitrobenzene. *Chemical Communications*, **46**, 7205-7207.

98. Zhu, C., Xuan, W., and Cui, Y. (2012) Luminescent microporous metal-metallosalen frameworks with the primitive cubic net. *Dalton Transactions*, **41**, 3928-3932.

99. Pramanik, S., Hu, Z., Zhang, X., Zheng, C., and Kelly, S., and Li, J. (2013) A systematic study of fluorescence-based detection of nitro-explosives and other aromatics in the vapor phase by microporous metal-organic frameworks. *Chemistry: A European Journal*, **19**, 15964-15971.

100. Rachuri, Y., Parmar, B., Bisht, K. K., and Suresh, E. (2015) Structural studies and detection of nitroaromatics by luminescent 2D coordination polymers with angular dicarboxylate ligands. *Inorganic Chemistry Frontiers*, **2**, 228-236.

101. Sanda, S., Parshamoni, S., Biswas S., and Konar, S. (2015) Highly selective detection of palladium and picric acid by a luminescent MOF: A dual functional fluorescent sensor. *Chemical Communications*, **51**, 6576-6579.

102. Parmar, B., Rachuri, Y., Bisht, K. K., and Suresh, E. (2016) Syntheses and structural analyses of new 3D isostructural Zn(II) and Cd(II) luminescent MOFs and their application towards detection of nitroaromatics in aqueous media. *Chemistry Select*, **1**, 6308-6315.

103. Rachuri, Y., Parmar, B., Bisht, K. K., and Suresh, E. (2017) Multiresponsive adenine-based luminescent Zn(II) coordination polymer for detection of Hg^{2+} and trinitrophenol in aqueous media. *Crystal Growth & Design*, **17**, 1363-1372.

104. Fu, H. R., Yan, L. B., Wu, N. T., Ma, L. F., and Zang, S. Q. (2018) Dual-emission MOF⊃dye sensor for ratiometric fluorescence recognition of RDX and detection of a broad class of nitro-compounds. *Journal of Materials Chemistry A*, **6**, 9183-9191.

105. Wu, P., Liu, Y., Li, Y., Jiang, M., Li, X., Shi, Y., and Wang, J. (2016) A

cadmium(II)-based metal-organic framework for selective trace detection of nitroaniline isomers and photocatalytic degradation of methylene blue in neutral aqueous solution. *Journal of Materials Chemistry A*, **4**, 16349-16355.

106. Xu, H., Chen, R., Sun, Q., Lai, W., Su, Q., Huang, W., and Liu, X., (2014) Recent progress in metal-organic complexes for optoelectronic applications. *Chemical Society Reviews*, **43**, 3259-3302.

107. Yi, F. Y., Wang, Y., Li, J. P., Wu, D., Lan, Y. Q., and Sun, Z. M. (2015) An ultrastable porous metal-organic framework luminescent switch towards aromatic compounds. *Materials Horizons*, **2**, 245-251.

108. Rachuri, Y., Bisht, K. K., Parmar, B., and Suresh, E. (2015) Luminescent MOFs comprising mixed tritopic linkers and Cd(II)/Zn(II) nodes for selective detection of organic nitro compounds and iodine capture. *Journal of Solid State Chemistry*, **223**, 23-31.

109. Majewski, M. B., Noh, H., Islamoglu, T., and Farha, O. K. (2018) NanoMOFs: Little crystallites for substantial applications. *Journal of Materials Chemistry A*, **6**, 7338-7350.

110. W. J. Rieter, K. M. L. Taylor, H. An, W. Lin, W. Lin, (2006) Nanoscale metal-organic frameworks as potential multimodal contrast enhancing agents. *Journal of the American Chemical Society*, **128**(28), 9024-9025.

111. Ni, Z., and Masel, R. I. (2006) Rapid production of metal-organic frameworks via microwave-assisted solvothermal synthesis. *Journal of the American Chemical Society*, **128**(38), 12394-12395.

112. Li, R., Yuan, Y. P., Qiu, L. G., Zhang, W., and Zhu, J. F. (2012) A Rational self-sacrificing template route to metal-organic framework nanotubes and reversible vapor-phase detection of nitroaromatic explosives. *Small*, **8**(2), 225-230.

113. Parmar, B., Rachuri, Y., Bisht, K. K., Laiya, R., and Suresh, E. (2017) Mechanochemical and conventional synthesis of Zn(II)/Cd(II) luminescent coordination polymers: Dual sensing probe for selective detection of chromate anions and TNP in aqueous phase. *Inorganic Chemistry*, **56**, 2627-2638.

114. Rachuri, Y., Parmar, B., and Suresh, E. (2018) Three-dimensional Co(II)/Cd(II) metal-organic frameworks: Luminescent Cd-MOF for detection and adsorption of 2,4,6-Trinitrophenol in the aqueous phase. *Crystal Growth & Design*, **18**, 3062-3072.

115. Song, J. F., Li, Y., Zhou, R. S., Hu, T. P., Wen, Y. L., Shao, J., and Cui, X. B. (2016) A novel 3D Cu(I) coordination polymer based on Cu_6Br_2 and $Cu_2(CN)_2$ SBUs: in situ ligand formation and use as a naked-eye colorimetric sensor for NB and 2-NT. *Dalton Transactions*, **45**, 545-551.

116. Li, A., Li, L., Lin, Z., Song, L., Wang, Z. H., Chen, Q., Yang, T., Zhou, X. H., Xiao, H. P., and Yin, X. J. (2015) Guest-induced reversible structural transitions and concomitant on/off luminescence switching of

an Eu(III) metal-organic framework and its application in detecting picric acid. *New Journal of Chemistry*, **39**, 2289-2295.

117. Singha, D. K., Bhattacharya, S., Majee, P., Mondal, S. K., Kumar M., and Mahata, P. (2014) Optical detection of submicromolar levels of nitro explosives by a submicron sized metal-organic phosphor material. *Journal of Materials Chemistry A*, **2**, 20908-20915.

118. Wu, Z. F., Tan, B., Feng, M. L., Lan, A. J., and Huang, X. Y. (2014) A magnesium MOF as a sensitive fluorescence sensor for CS2 and nitroaromatic compounds. *Journal of Materials Chemistry A*, **2**, 6426-6431.

119. Barman, S., Garg, J. A., Blacque, O., Venkatesan, K., and Berke, H. (2012) Triptycene based luminescent metal-organic gels for chemosensing. *Chemical Communications*, **48**, 11127-11129.

120. Zhang, M., Zhang, L., Xiao, Z., Zhang, Q., Wang, R., Dai, F., and Sun, D. (2016) Pentiptycene-based luminescent Cu (II) MOF exhibiting selective gas adsorption and unprecedentedly high-sensitivity detection of nitroaromatic compounds (NACs). *Scientific Reports*, **6**, 20672.

121. Nagarkar, S. S., Desai, A. V., and Ghosh, S. K. (2014) A fluorescent metal-organic framework for highly selective detection of nitro explosives in the aqueous phase. *Chemical Communications*, **50**, 8915-8918.

122. Nagarkar, S. S., Desai, A. V., Samanta P., and Ghosh, S. K., (2015) Aqueous phase selective detection of 2,4,6-trinitrophenol using a fluorescent metal-organic framework with a pendant recognition site. *Dalton Transactions*, **44**, 15175-15180.

123. Yang, J., Wang, Z., Hu, K., Li, Y., Feng, J., Shi, J., and Gu, J. (2015) Rapid and specific aqueous-phase detection of nitroaromatic explosives with inherent porphyrin recognition sites in metal-organic frameworks. *ACS Applied Materials & Interfaces*, **7**, 11956-11964.

124. Kim, T. K., Lee, J. H., Moon, D., and Moon, H. R., (2013) Luminescent Li-based metal–organic framework tailored for the selective detection of explosive nitroaromatic compounds: Direct observation of interaction sites. *Inorganic Chemistry*, **52**, 589-595.

125. Zhang, Y., Li, B., Ma, H., Zhang, L., and Zhang, W. (2017) An RGH-MOF as a naked eye colorimetric fluorescent sensor for picric acid recognition. *Journal of Materials Chemistry C*, **5**, 4661-4669.

Index

C

D

E

F

G

U

UV-Vis spectrophotometer,
219

V

van der Waals forces,
13-14, 175
viscosity, 41, 46-47
volatile organic solvents,
225
volume fraction, 37

W

water treatment, 1, 7, 28,
73, 144
wave function, 152-154

Z

zeolite imidazolate
framework, 10, 126